JOE MUCCIA

Jump into the Fight

The Complete, Unabridged History of Easy Company,
Volume I

First published by MHRP, LLC 2026

First edition

ISBN: 979-8-218-39351-9

Cover art by Krysti Muccia

This book was professionally typeset on Reedsy.
Find out more at reedsy.com

For Gary, Marianne, Syndee, Karen and Tina
I'm sorry I couldn't get it done in time...

Contents

Acknowledgments

First, I'd like to thank the Easy veterans and their families, as well as those of the broader 506th family. Without their contributions, this book and its subsequent volumes would not have been possible.

Originally, I had no desire to "write what has been written." Ambrose got the ball rolling. After him, it has been Marcus Brotherton who has wonderfully done the lion's share in capturing a larger portion of Easy Company's history. But many years ago, I stood on the tarmac at Reading Airport in Pennsylvania talking with Easy veteran Ed Tipper. We had been discussing Easy's history and with his typical directness, he jabbed his pointer finger into my chest several times. "You should write a book about the Company. I don't think I've ever met anyone who knows as much about us as you do. You could make a real contribution to the unit history." Ever the schoolteacher, Ed was instructing me to get on with it.

Still, I mulled the idea over for many years and the more I thought about it, the more I realized how many stories still hadn't been told. After corresponding with Easy veteran Les Hashey for a while, he sent me a copy of his wartime photos with a note that said, "Do a good job taking care of our history. It's in your hands now." Hesitantly, I ran the idea past the

Easy families. To me they have always held, "the right of first refusal," about anything Easy-related that I intended to publish. My initial thought was to write and publish it privately for them. But over the years, Chris Langlois (one of Doc Roe's grandsons) continued to push me to publish it for the masses. He finally wore me down. So, here we are.

In addition to the veterans and their families, I'd like to also thank the following for their contributions to this series. This would not have been possible without their assistance: Doug Barber, Marcus Brotherton, David Campbell, Steve Cambetes, Brenda Carlan, Tony Coulter, Ian Gardner, Kenny Gunther, Leighton Hughes, Reg Jans, Erik Jendresen, Dan Kveton, Sean Lester, The Makos Family, Marty Morgan, Phillip Russo and Sue Hardy VerHoef. Apologies to anyone that I've omitted. It takes a village to raise a child, and it took a village to make this series happen.

Joe Muccia

Fredericksburg, VA

February 2026

Introduction

Despite all the criticism he has received in both public and private circles, Stephen Ambrose had it right when it came to why Company E, 506th Parachute Infantry Regiment of the 101st Airborne Division, was special. They came from all walks of life and every type of background. Some were drafted. Some were motivated by patriotism in the wake of Pearl Harbor. Others just wanted to belong to something unique. Some of them wanted a say in their own destiny instead of being tossed into the assembly line-like machinery of the wartime U.S. military.

Regardless of the reasoning or background, these men were thrown together and forged into a fighting team, beginning with the crucible of Toccoa and then further tempered in the fire of combat. In surviving that experience, they became a group of men so close that through 2012, they had gathered for reunions each and every year. While some historians consider them to be an average company in an outstanding combat division, it was not just their wartime service that made them special, but also the enduring legacy of family and friendship that was built on top of the foundation of that service in World War II.

The purpose of this book is not to recreate the wheel with regard to the history of Easy Company. It is simply to illuminate further, the story of the men that made up Easy. Much has been made about the actions of the company in battle, but those actions were not the primary drivers for their immense popularity. Ambrose's book and the subsequent miniseries of the same name told the story of men, men that people could relate to and identify with. Almost every one of the men who served in Easy could be identified with as someone we all knew in our own lives, whether it be an uncle, brother, father or grandfather. It was a story of ordinary men accomplishing extraordinary feats. And while many of them knew they were combating evil incarnate, most did their duty for the man to the left and to the right of them.

Ambrose humanized these almost mythical warriors and that is why people have been so drawn to them over the years. The mythical has been personified and made accessible. So, with that in mind, I decided to collect their stories, many of which have never been heard, and bring them together in this series of books. Hopefully, it will further enrich the already robust legacy of Easy Company.

This first volume covers the formation of Easy Company and its subsequent training in the United States. It details all the training the men endured and does so in their own words, based on my interviews with them, their own written words, and from previously published materials. By weaving these sources into one narrative, it hopefully will provide the reader with the most complete accounting of Easy Company's history, from the first bugle call at Toccoa, to when the men walked up the gangplank and onto the *RMS Samaria*. Subsequent volumes will cover the unit's service in Normandy, Holland, and Bastogne until the end of the war. It is my hope that it will satisfy even the most ardent fan's desire for information about this Band of Brothers. With that said...

This series is dedicated to the Men of Easy Company and their families.

1

From Every Corner of the Nation

"When you went through Toccoa, you went through the hardest the Army had to offer."
– Don Malarkey

Camp Toccoa, August 1942 - November 1942

As with almost every story about Easy Company, this one begins at Camp Toccoa, or Camp Toombs as it was first known. The camp was originally built in the 1930's to support President Franklin D. Roosevelt's Works Projects Administration.[1] These projects were created by the U.S. Government to help alleviate the economic crisis by providing jobs for young men during the Great Depression. Toombs was meant to support the training of National Guard units, but that mission changed dramatically in the wake of the Japanese bombing of Pearl Harbor. America was at war and its society was being mobilized to support that effort. As a result, Toombs became the initial training center for many U.S. paratroopers-to-be.

Originally called Camp Toombs, in honor of Confederate

Civil War General Robert Toombs, the name of the camp was changed to Toccoa after the commander of the 506th Parachute Infantry Regiment, Colonel Robert F. Sink, became aware that arriving troopers had to pass the Toccoa Casket Company on their way to the main camp. Sink said, "Things have to get better if they are this bad in the beginning," referring to seeing the casket factory which he feared would have an adverse impact to the unit's morale.[2]

Robert Strayer, the soon-to-be commander of the Regiment's Second Battalion, had this to say about his early arrival at Camp Toombs, and the name change. "Bob Sink and I were the first officers to arrive at Camp Toombs. I was a captain then. When Colonel Sink found out there was a casket factory nearby, he immediately petitioned the War Department to have the name changed to Camp Toccoa."[3] A young Second Lieutenant, Richard D. Winters, recalled the casket company upon his arrival in the sleepy Georgia town. "On the way to the camp I was pretty unsettled. I took Highway 13, passed a casket factory and reported in at Camp Toombs."[4] So, while news of the camp's name hadn't spread very far outside of the general area, what had spread was news about a relatively new type of fighting man in the U.S. military, the paratrooper.

The three men that would help shape Easy Company from 1942 to 1943.
Left: Robert Strayer, commanded the Second Battalion of the 506th Parachute Infantry Regiment, from its formation in 1942 until his promotion to Regimental Executive Officer in 1945. (Photo courtesy of the Strayer Family)
Middle: Herbert Sobel, erroneously reported as the first commander of Easy Company through the years, through force of sheer will and personality, molded Easy Company into a tight-knit, "Band of Brothers." (Photo courtesy of Michael Sobel)
Right: Richard Winters, intelligent, physically fit and morally guided, Winters stressed fitness and mastery of soldierly skills to the young men of Easy Company. (Author's Collection)

Company E of the 506th Parachute Infantry Regiment was initially formed on 20 July 1942, according to U.S. Army General Order #23. At the time, it consisted of little less than a squad of troopers, and until the middle of August 1942, was primarily made up of prospective officers and cadre members. Outside of Wally Moore, who was initially assigned as the

Company Commander, the officers were either transferred to Regimental Headquarters or out of the paratroops altogether.

The term "paratrooper" was quickly becoming well known across America. Serials were shown in advance of movies at local cinemas. Frank Perconte, a 24-year-old Italian-American from Chicago, was one of those who were clearly influenced by those early attempts to spread the word. "Herman (Hanson) and I were at the Rialto Theater one day when they were showing a clipping of paratroopers and (it) was talking about how they paid extra money, so this is what made us join," Perconte said. "We were about to be drafted and didn't know when we would be going, but knew if we volunteered to become a paratrooper we were going to receive $100 a month," he said.[5] [6]

While many of the recruits who showed up were in their twenties, several were extremely young. Dewitt Lowrey was one of the youngest. "I saw Uncle Sam pointing at me and I saw a paratrooper sign and said, 'That looks like fun', so I decided I'd go with the paratroops. So, I went and joined the paratroopers. The recruiting sergeant carried me down back home so momma and daddy could sign the papers and I could go in. Momma didn't want to sign at first. Daddy told her, 'You might as well. They're gonna draft him anyway in a year or two, and this way he gonna get into what he wants to get into.' And so, they both signed the papers for me to get into the paratroopers. I was 17 or 18."[7] Lowrey was born in April 1922 and was nineteen at his time of enlistment...young for a soldier any way you slice it.

Amos "Buck" Taylor, of Philadelphia, also saw a movie reel describing the toughness of the paratroops and he was taken by its possibilities. "I was in a movie theater in my hometown of Philadelphia. They were showing a Fox Movietone News clip about the Army. It was about the paratroopers, and it looked interesting. I was always an outdoor person. I spent a lot of time hiking in the woods as a youngster. I thought to myself, this looks like something I would be interested in as opposed to going into the infantry and sitting in a trench somewhere or in a foxhole, so I looked into it and enlisted in July 1942."[8]

Left: Carwood Lipton and Frank Perconte, seen here after graduating from The Parachute School at Fort Benning. The pair were two of the first men to arrive and remain with Easy Company. (Photo courtesy of Frank Perconte)
Middle: Herman Hanson, who went by the nickname "Hack," joined up with his friend Frank Perconte. Hack is seen here in a photo taken at Toccoa. Notice despite the paratroop patch on his overseas cap, he hasn't qualified as a paratrooper yet and as a result, doesn't wear jump wings over his pocket. (Photo courtesy of the Hanson Family)
Right: James "Moe" Alley. Despite Sobel's edict that the men of Easy would not be permitted to wear their jump boots until qualifying as paratroopers, Moe wears his for a "hero" photo for the folks back home while standing on the PT field at Toccoa. (Author's Collection)

Carwood Lipton, a conscientious and mature young man from Huntington, West Virginia, had also heard about the paratroops, but from a different source. "Life magazine ran an article on paratroopers and how difficult it was to join them. Since I was in good physical condition at the time, I decided I would be one. I didn't tell anyone what I was doing until I had done it, not until after I enlisted and had to take a train to Kentucky the next morning. That way no one had an opportunity to second-guess me."[9] [10] Lipton was also patriotic, and after Pearl Harbor, he was well motivated, "The basic reason (I joined) was that my country was at war and I wanted to do what I could to help."[11] While at the reception center at Fort Thomas, Kentucky, Lipton met a boyish looking prankster named Wayne A. "Skinny" Sisk of Mullens, West Virginia, and Daniel B. "Boone" West,[12] a laconic, dark-haired young man from McCreary County, Kentucky. The trio would eventually make their way to Toccoa and Easy Company's Third Platoon.

Unlike Lipton, James "Moe" Alley, of Mount Ida, Arkansas, was second-guessed from the start. A veritable old man at 21, Alley joined after seeing a recruiting poster for paratroopers in his hometown. His parents, none too pleased by his decision, told him that he had signed his 'death warrant.' Still, the lanky Alley would not be swayed from his chosen path.[13] After signing the enlistment papers, Alley was shipped to Camp Joseph T. Robertson, where he sat, "around camp for what seemed like weeks," until he was finally called up to the camp headquarters. He was furnished a train ticket, and after a two-day trip, arrived at Toccoa. He would eventually be assigned to Easy's Third Platoon.[14]

Left: Daniel B. West. The soft-spoken West was as solid as they come. With first and middle initials of "D. B." the other members of Third Platoon often called him "Daniel Boone." (Photo courtesy of Cathy West Landis)
Middle: The boyish-looking prankster, Wayne "Skinny" Sisk, was repeatedly in Captain Sobel's doghouse, but no matter how badly he fouled up, Sobel never transferred him out of Easy Company. (Author's Collection)
Right: At nineteen, Dewitt "Alabama" Lowrey was one of the youngest soldiers to join Easy at Toccoa. (Author's Collection)

Perconte and Hanson, along with Lipton, Sisk, and West were part of the first 43 men to arrive at Toccoa and be assigned to Easy Company. Outside of those four, of the 43, only 13 would spend any significant amount of time with Easy.[15] One of the first people to greet them upon their arrival at Toccoa was First Lieutenant Herbert Maxwell Sobel. A Jewish man from a working-class neighborhood in Chicago, Sobel was the second officer to be permanently assigned to Easy Company. Once he arrived on 11 August 1942, Sobel was appointed the company's commanding officer by Colonel Sink.

Upon their arrival from W Company, the five new men reported to First Lieutenant Sobel. Once that was completed, a corporal marched them down to their respective barracks. Lipton recalled, "We were introduced to the 1st Sergeant, Harvey Morehead, and our platoon sergeant, Staff Sergeant Kudla, both regular Army cadre, and were told that we were the first men from W (Company) to come into E Company. We were put in the first barracks, next to the orderly room and told that we were the 1st Platoon. As the company began filling, though, our platoon was renamed the 3rd Platoon so that when we came out of the barracks for company formations we would be in the right position as the company formed up, without having to go to the far end of the company street."[16] Perconte eventually ended up in First Platoon, Hanson to Second, while Lipton, Sisk and West remained in Third Platoon.

Some of the other arriving men, like Donald "Hoobs" Hoobler, Robert "Rook" Rader, William "Shep" Howell and Alex "Pink" Penkala, had served in the peacetime National Guard. All four became enamored with the thought of being a paratrooper but in order to do so, they had to be discharged from the National Guard and reenlist in the regular Army. Once that was complete, they had to volunteer for the paratroopers. According to Penkala's family, several of his friends also volunteered for the paratroops. While they washed out, Alex eventually made the cut. He had trained at Fort Lee to be a cook but obviously wanted something more exciting.[17] Each of them arrived in late August and by September 1st, 1942, the four had been assigned to Easy Company.

Ed "Tip" Tipper, the son of Irish immigrants living in Detroit, had graduated from high school and had worked two different factory jobs to help his parents make ends meet. Having left those two jobs, he was delivering pianos for a department store when the war broke out. By August of 1942, he had grown bored of that tedious, backbreaking job and decided to join the paratroops. On the train from Detroit to Toccoa, he met Walter "Wally" Wentzel. The two would maintain a lifelong friendship and when Walter's son Keith was born, he asked Ed to be his godfather. But that would take place years in the future. For now, the pair, with their destinies entwined, endured the long, slow trip south.[18]

The attack on Pearl Harbor was still fresh in everyone's mind in mid-1942 and it enraged many of the future Easy men. Twenty-year old Johnny Martin, a Columbus, Ohio native, was one of those motivated to enlist by the attack. "My reason to enlist was that I was working in the railroad and the Japanese had just bombed Pearl Harbour (sic). The Japanese were slaughtering our Marines in Corregidor and Bataan. I decided to enlist at Fort Benjamin in Harrison, Indiana. The recruiter asked me for my experience, which was in the area I had in the railroad. He suggested railroad work, but I said 'No.' I wanted somewhere I could be more effective as soon as possible. He suggested the airborne." Martin fell for the Airborne 'pitch' hook, line, and sinker. "Mark me down for the parachute troops," he told the recruiter at Fort Benjamin Harrison. Martin soon found himself headed to Toccoa, Georgia.[19]

Left: Joe Toye of Reading, Pennsylvania, was tough-as-nails, but had a heart of gold. An experienced soldier who had served with the 29th Infantry Division, Toye quickly climbed the NCO ranks of Easy Company. (Author's Collection)
Middle: Darrell Powers, a soft-spoken southern gentleman from Clinchco, Virginia, "Shifty" had earned his nickname on the basketball court. Already a crack shot from hunting as a youth, Powers would become one of Easy's deadliest marksmen. (Author's Collection)
Right: Robert "Popeye" Wynn, another Virginian, joined the paratroops on a $5 bet with his good friend "Shifty." (Author's Collection)

Joe Toye too was spurred on to enlist by the attack on Pearl Harbor as well, but he had taken a convoluted route to get to Toccoa. Joe enlisted at Wilkes-Barre, PA on Dec. 11, 1941, just four days after the Japanese attack. After basic training, he was stationed in Washington, D.C. with the 29th Infantry Division, as a medic and truck driver.[20] The posting soon bored him, as he got tired of being part of dog-and-pony shows for the brass. He yearned to find a way out of the humdrum duty and found that the only way he'd be granted a transfer was if he volunteered for the Airborne. Enticed by the lure of jump pay and possible combat duty, Toye volunteered for the paratroopers and was soon on his way to Toccoa and Easy's Second Platoon.[21]

Leo Boyle, from Roseburg, Oregon, served in the peacetime Army beginning in 1934 as an ordnance specialist. After being stationed in Honolulu and Fort Lewis, Washington, Boyle was discharged in 1937. But the attack on Pearl Harbor spurred something in him and he decided to sign up again. Like Joe Toye, he had no desire to serve in some kind of support role knowing it would be monotonous and full of drudgery. "I jumped at the idea of getting into some kind of special service where there'd be more action and more movement." He knew that signing up for the paratroops would be "not only exciting but dangerous." After arriving at Toccoa and being processed through W Company, Boyle was assigned to the fledgling company's headquarters section.[22]

Pearl Harbor had another effect on prospective servicemen. Nineteen-year-old Darrell Powers, a native of Clinchco, Virginia, had been working since he left high school. While in high school, Powers played basketball, and because of his agility on the court, was given the nickname "Shifty." He had a good job in the defense industry that provided a draft deferment, but that deferment had its own consequences. "After the Japanese bombed Pearl Harbor, they (the government) moved all of us from the school over to the Navy Shipyard at Portsmouth, to do work on the ships there. I worked there as a machinist for a while, I liked the work. But then we heard that the government was getting ready to freeze our jobs there meaning we'd have to stay there until the war was over. My buddy (Robert) "Popeye" (Wynn) and me wanted to join the military so we went and signed up for the Army before we got stuck at the shipyard. We volunteered for paratrooper school."[23] What "Shifty" hadn't mentioned was that he and "Popeye" had a $5 bet on who would chicken out first. Neither did and after initial processing at Camp Pickett, both Virginians were soon shipped out by train to Toccoa and Easy's Third Platoon.[24]

While Sobel was the chief trainer for the company, he was not alone in this endeavor. A group of Old Army NCOs assisted him with one assigned to each rifle platoon as well as the company headquarters. Those cadre members were First Sergeant (1/SGT) Harvey Morehead (Company HQ), Staff Sergeant (S/SGT) Joe Stedman (First Platoon), Staff Sergeant (S/SGT) Norman Ford (Second Platoon), and Staff Sergeant (S/SGT) Steven Kudla (Third Platoon). The men recalled the cadre as being "pretty hard-boiled." They were salty Old Army NCOs, tough and exacting, with a colorful grasp of barracks-style profanity. According to Walter Gordon, "We looked up to them almost like gods because they had been to Panama, they had their wings, they were qualified jumpers and we just thought they were the last word."[25]

Ed Tipper, who started out in First Platoon before eventually moving to Third, experienced the respect that the cadre men garnered firsthand early on in his tenure with Easy. "At Toccoa we had a no-nonsense drill sergeant named Harvey Morehead. He was tough as nails and played a strong role in having me grow up. One time we were standing at attention and I made some comment to the guy next to me. Sergeant Morehead barked out a command: "You men are at attention. There is no talking in ranks. And Private Tipper—" he spoke the next phrase very deliberately, "—one more word from you and you are out!" I knew he meant it. I chose to shape up right then."[26] While Morehead could threaten the men with punishment, Stedman in particular, was tough and intimidating in his own right. It was rumored that Stedman had been the middleweight boxing champion in the fledgling paratroopers at one point. Even if it wasn't true, few of the men could physically challenge him. This overt threat helped to keep the young paratroopers-to-be in line.

Left: Walter Gordon attempted to use reverse logic on the Army in order to get stationed closer to his home of Louisiana. It only worked in part due to his desire to become a paratrooper. (Photo courtesy of the Gordon Family)
Middle: Harvey Morehead was the second first sergeant assigned to Easy Company (the first was Albert Miller who eventually went to Regimental Headquarters). He led the other three Old Army cadre men that initially trained Easy Company to be soldiers. (Author's Collection)
Right: Ed Tipper learned a lot about soldiering early on in his tenure with Easy Company, the first item being instantaneous adherence to orders. Harvey Morehead was a critical factor during that period. (Author's Collection)

As the days on the calendar ticked off, more Easy Company troopers made their way to Toccoa, from early August and into September of 1942, most if not all, arrived by rail. Robert Rader recalled that he, the mischievous Don Hoobler, and easy-going "Shep" Howell, formed the "Ohio Contingent" of the first recruits, along with a later-arriving Johnny Martin.

According to Rader, it had been Hoobler's, "bright idea to jump out of airplanes." After settling in, Rader fondly remembered the good-natured ribbing the three took from the other recruits because of their regional dialect and accents. "Shep" Howell continuously ran afoul of First Sergeant Morehead as "Shep" pronounced his last name like "Owl". As a result, Morehead mustered him as absent for his first week in Easy. "God dammit, if he don't (sic) know his name, I'll send him to school to learn how to say it!" Morehead roared. The issue was soon resolved but not until after the three Ohioans bore the brunt of some ribbing from the cadre and their fellow troopers due to the "funny way they talked." All three men ended up being assigned to Easy's First Platoon.[27]

Three members of the "Ohio Contingent." Robert "Rook" Rader, Donald "Hoobs" Hoobler, and William "Shep" Howell, seen here while Easy was at Clemson University qualifying with their weapons. (Photo courtesy of the Carson Family)

For others, like Rod Bain, boot camp in Georgia marked the first time he'd traveled outside the Pacific Northwest. "It was a shock," he said of that time. In fact, it was Rod who helped to welcome Don Malarkey into the ranks of Easy, along with Tom Burgess of Humboldt, Illinois.[28] Bain and Burgess bunked next to each other in the barracks. Malarkey was a good-looking Oregonian with a curly mop of blonde hair. As he set up his cot, Malarkey noticed a picture of a girl on Bain's shelf. The photographer's stamp placed the photo in Astoria, Oregon, Malarkey's hometown. He figured he knew every girl in Astoria, but he'd never seen this one. "Who's that girl?" he asked the pair as they entered the barracks. Bain, from across the Columbia River in Ilwaco, Washington, spoke up, saying she was his sister. The two chatted. From then on, they were friends. "I suddenly felt I'd never even left home," Malarkey said.[29]

For Malarkey he remembered other details of his arrival and indoctrination vividly, including arriving on the same train as the "Ohio Contingent." "When I arrived in Toccoa, it was the first week of September 1942. I came to Toccoa with a shipment of guys out of Ohio. A soldier named Bob Rader, and another named Don Hoobler. The three of us ended up at Toccoa at the same time. I was put in a tent with five or six guys who were getting kicked out of Toccoa. For a day and a half all I heard was them complaining that "they didn't give me a chance, they kicked me out," and such, and they kept talking about the mountain, which scared the hell out of me. So, they took me up to E Company, and I set up a cot at the end of the barracks."[30]

After he settled in, Bain and Burgess took Don for his first dinner at the mess hall as a member of Easy. After washing their mess kits, the three made their way over to the camp's meager PX for a soda before heading back to the barracks. On their way back, "...in the early twilight, they stopped and pointed out Mt. Currahee, telling me that it was 'my' problem, as running it several times a week was part of the training. Falling out of a run was a 'NO-NO,' and qualified the failure to become a candidate for the "moonlight patrol", triggered by a tap on your shoulder in the middle of the night with a command performance up the mountain. If it was their intention to scare me, they had been successful. However, they agreed to take me up the mountain the next night so that I could be sure I could run it."[31] With that in mind, the trio retired to the barracks of Easy's fledgling Second Platoon.

Mount Currahee looms large from the trail leading up to the mountain. Bill Wingett recalled rocks the size of softballs littering the trail, some of which can be see in this picture. (Photo courtesy of Denver "Bull" Randleman via Chris Langlois)

One of the most interesting enlistment stories was that of Walter "Smokey" Gordon. Walter had been attending courses at Millsaps College when the attack on Pearl Harbor happened. Like many others, he tried to enlist but was turned down due to color blindness. Walter then traveled north to Philadelphia to enter the service. The reason for doing so was twofold. He thought he could sidestep the color blindness issue, and he also believed that he would be stationed far from his home in Louisiana if he enlisted in the south. Based on his idea of Army logic, he figured the Army would station him in the south if he enlisted up north.His plan would eventually be thwarted by his own desire to become a paratrooper. Because he volunteered for the Airborne, the medical examination at the induction center wasn't quite as stringent as the ones he had previously endured. This was likely because of the Army's desire to fill up the ranks of still-forming paratrooper units. "...at the risk of sounding corny, I felt that all young people at that time had a general feeling that wanting to participate and to do something significant other than to drive a truck, or be a mess cook...(and) there's nothing wrong with those two duties, but I preferred to do something a little bit more meaningful in trying to resolve the conflict. So, I asked for it and felt the Airborne would serve that."[32]

But at least Gordon wouldn't have to travel to Toccoa alone. He was joined on his train ride by Forrest "Goodie" Guth, Roderick Strohl, Carl Fenstermaker, and Amos "Buck" Taylor. Gordon referred to Guth, Strohl, and Fenstermaker as the "Pennsylvania Dutchmen." Goodie recalled those early days well. "(I) was exempt from the draft because (I) had a job at Bethlehem Steel. Still, (me) and two friends (Strohl and Fenstermaker, all from Fogelsville, PA) joined the army and volunteered for the airborne because paratroopers made $50 a month more than regular army pay of $27 a month."[33] As the trio sat there in the recruiting office, they dared each other to join the paratroops. "If you will, I will." The dares worked. All three ended up volunteering together and even ended up with sequential Army Service Numbers. After a week long stop at Fort Indiantown Gap, Pennsylvania, they were put on a train to Georgia.[34] With Gordon in charge of the group for the train ride, they finally arrived at Toccoa two days later. Gordon, the 'Dutchmen', and Taylor, were processed through W Company, and eventually all were assigned to Easy's Third Platoon.

Left: Amos "Buck" Taylor was a natural at physical training and he excelled at that aspect of soldiering. "Buck" believed that was part of the reason why he was promoted early on in his tenure with Easy Company. (Photo courtesy of Amos Taylor)
Middle: Forrest "Goodie" Guth, one of the three Pennsylvania Dutchmen. Goodie hated running Mount Currahee and if he had his way he would have abolished it from the training schedule. A self-described chow hound, Goodie grew to hate SOS which was almost always on the menu at the chow hall. (Photo courtesy of Forrest Guth)
Right: Roderick Bain, along with Tom Burgess, welcomed Don Malarkey into Easy by running him up Mount Currahee on his second day in the company. (Photo courtesy of the Bain Family)

"Buck" Taylor's recollections of that time were also particularly strong. "In August (1942) I went up to the inductee camp and met Walter Gordon, who was there. He was from New Orleans, but he signed up in Philadelphia for some reason. He was up there along with Forrest Guth and Rod Strohl, at Indiantown Gap (Pa.). We were only there for two or three days, and then we went on the train on down to Toccoa. And that started our big experience. Gordon, Guth, Strohl and I all arrived by train at the same time. We wound up in the same platoon."[35]

Prospective troopers from the West Coast began arriving at the camp as well. Richard "Red" Wright took the train as part of a group led by future Third Platoon Sergeant, Terrence "Salty" Harris. Both would eventually end up in the 506th Pathfinders. Bill Wingett also came by rail, along with Burton "Pat" Christenson, William "Duke" Dukeman and Everett "EJ" Gray. Wingett recalled the experience with his new buddies. "I shipped out the next morning, Aug 20, 1942, to Atlanta by commercial train. I had some companions for the trip, Burton Christenson from Oakland and William Dukeman and Everett Gray. Dukeman and Gray were high school buddies from Keensburg, Colorado, and happened to be in San Francisco because they had been on a motorcycle trip. Their bike had broken down or had mechanical problems, so they decided to end the trip, sell the bike and enlist. We had all read a brochure about airborne troops in the recruiting station, and we decided that might be an exciting route to go. Our Army serial numbers were only a few digits apart. I was assigned to the Second Squad of the First Platoon. Christenson, Dukeman and Gray stayed in my platoon, and they became my best friends. Christenson and Dukeman were in the First Squad; Gray was in my squad."[36]

Left: Robert Burr Smith, or "Sergeant West Point," was a tall, good-looking Californian who was skilled in manual-of-arms with the M1 Garand service rifle. (Author's Collection)
Middle: Burton "Pat" Christenson, along with Burr Smith, was another of what Dick Winters called, Easy Company's "California Golden Boys." He took his physical training seriously and always ranked at the top of the company's scorecard. (Photo courtesy of the Christenson Family)
Right: Coloradan William Dukeman joined the paratroops with his childhood friend, Everett Gray, after the pair's motorcycle broke down in California. (Photo courtesy of the Carson Family)

As the new Easy Company troopers arrived at Toccoa, what they found wasn't very inspiring and many even questioned their decision to become paratroopers based on what they saw. The camp was spartan at best. In many instances, there were no paved roads. When it rained, paths and trails became quagmires. Many of the tarpaper barracks buildings were still being constructed and unlike other camps, it didn't even possess a recreation hall or movie theater. Roofs leaked and cots sunk into the soft Georgia clay. In short, the camp was a partially built mess.

Left: Tarpaper barracks are clearly visible in this photo as are the brick buildings that comprised much of the Regimental Staff area. (Photo courtesy of Denver "Bull" Randleman via Chris Langlois)
A view down one of the company streets at Camp Toccoa. Tarpaper barracks can be seen on the left. (Photo courtesy of Richard Winters via Erik Jendresen)

Lieutenant Winters remembered the condition of the camp and the sparseness of it. "There was not much there, and I was assigned to a tar paper shack. There were no windows in any of the buildings, and the only place with electricity was the latrine. This was rough. But you were expecting to have it rough if you were going to be in the parachute troops." Walter Gordon echoed Winters' assessment; "...it was a dismal little place, sitting on some sandy hills and it had been a former

National Guard camp...it was just stuck out there in the middle of nowhere."[37]

"Buck" Taylor saw Toccoa in similar terms. "Toccoa was just a mud hole. It was a camp that was just being put together, and it was all a mess of red clay. It really didn't have any facilities yet."[38] Private Robert T. "RT" Smith had a similar impression of the camp after one week of training. "This week it has been raining off and on every day and so it is pretty bad wallowing around in this red Georgia clay, that is all they have here and we get plenty muddy."[39] While the Regiment's First and Second Battalions were at least assigned to tar paper barracks, troopers in Third Battalion had it worse...they were 'housed' in squad tents.

Burr Smith, a Californian who was eventually assigned to company headquarters, began rubbing others the wrong way when he arrived. According to Robert Rader, "His personality was such as to lead you to believe that he had done everything. He had a 'nickel's worth' to say about any subject mentioned."[40] But over time he won over the other men, including Rader, with his good looks and charm. Smith, who had arrived with childhood friend Richard "Dick" Berg, recalled the bleakness of the camp in its earliest days, especially the Regimental processing camp, uniquely named, W Company. Upon arrival at the camp, the recruits were assigned to W Company for in-processing. This typically took between three and five days.

W Company was where prospective paratroopers both came and went, sometimes on the same day, and in September of 1942, "W" was nothing more than a tent city on the grassy slope of a hill just below the regimental medical processing facility. The neophyte troopers referred to W Company as

"Washout Company" or "Cow Company," due to the many outgoing troopers that milled about the area. The squad tents, as brand new as the citizen soldiers who occupied them, were aligned to form a company street, but W Company was a company in name only. "It served as the Regiment's in-and-out processing machine, and it was a fast train in both directions. The incoming volunteers (mostly draftees, some enlistees, but all volunteers for parachute training) were frantically busy from morning to night...drawing clothing and equipment, filling out forms, falling in for meals, marching to examinations, etc. The train was moving much too fast to jump from it and there was never, to my knowledge, a single disciplinary action among the thousands of 'in-processes'," recalled Burr Smith.[41]

During in-processing, potential recruits were put through a battery of medical examinations and written and physical tests to determine suitability for duty in the Airborne. Little has been mentioned of this in the past, but thankfully "Shifty" Powers remembered those early examinations in some detail. "Even though we had had complete physicals already, when we got to Toccoa, we had to go through it all over again. And these were a little different," Powers said. "They had different regulations for paratroopers. You had to be so tall but not too tall. So big but not too big (weight-wise). You had to have good hearing. You couldn't have had any broken bones, even though they had healed. That eliminated a lot of boys that had played football in school. You pretty much had to be in prime condition."[42] Johnny Martin was a little more pointed in his opinion of the testing. "I took some examinations, basic examinations, physical and a kind of half-assed IQ test, which I really didn't give a damn about impressing anybody with."[43]

After a brief stay in W Company, he was assigned to Easy's First Platoon. Ever sharp, Martin eventually became one of Easy's earliest non-commissioned officers, or NCOs.

After finishing his evaluations at W Company, Carwood Lipton spotted the notes the doctor had scribbled on his personnel file, "Excellent on calisthenics and coordination. Pes Planus, 1st degree." The latter meaning he had flat feet, although it did not disqualify him from duty.[44] At one point, Pat Christenson and Robert Rader were going through the same medical processing and listened to the exchange between a prospective trooper and a medical officer. "Son, you're just not built for this kind of outfit." Hearing this, Christenson thrust his chest out as the medical officer moved towards him. He and Rader were both accepted.[45]

For Bill Guarnere entering the service wasn't that foreign of an experience. He had been part of the Civilian Military Training Camp program, an effort to get kids off the street, giving them some exposure to the military, and a way for them to make a little bit of money. He had come from the tough streets of South Philly and by the time he reached W Company, he was ready for Army service, especially the "looks and the challenge of the Airborne." He wasn't impressed with what he saw while at W Company, especially with the mud that seemed to coat everything. With his South Philly chip on his shoulder, it didn't take the combative Guarnere long to find a willing "dance" partner. He and Richard Bray ended up trading punches until they were broken up by their tent mates. Both would end up assigned to Easy Company, Guarnere in Second Platoon, and Bray in First Platoon. Upon arriving at E Company, Guarnere was quickly befriended by Johnny Martin, and it was Martin that would eventually bestow on Guarnere his unforgettable nickname, "Gonorrhea."[46] [47]

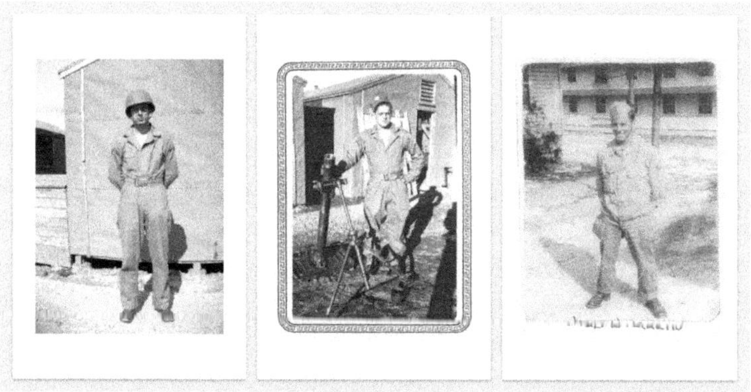

Left: William Guarnere, seen here in his coveralls at Camp Toccoa, brought innate leadership qualities and a combative nature that would serve him well throughout his military service. (Photo courtesy the Guarnere Family)
Middle: Don Malarkey was welcomed to Easy Company by Rod Bain and Tom Burgess. Seen here wearing newly-issued coveralls, Don poses with an 81mm mortar from the 2nd Battalion Headquarters while at Toccoa. (Photo courtesy of Don Malarkey)
Right: Johnny Martin tried to join the Marines first (as did Don Malarkey) but he didn't want to wait to get into the fight. He felt the paratroops gave him the best opportunity to do so. (Photo courtesy of Phil DiMaria)

While still assigned to W Company, Don Malarkey soon saw a familiar face after finishing a meal at the mess tent. Once a recruit ate his meal, he was required to wash his mess kit in a steel garbage can filled with boiling hot water and present the mess kit for inspection by a Regimental officer. As Don moved up the line, he dunked his mess kit as ordered. As he picked up his head to meet the gaze of the inspecting officer, he recognized someone from his time at the University of Oregon, a fellow by the name of Gene Brown. The pair had crossed paths several times while at school and knew each other. Brown had joined the paratroops and had been assigned to the 506th as a second lieutenant. Pleased to see someone he knew, Don blurted out, "Hi, Gene." Brown looking slightly flustered responded in kind, "Hello, Don."[48] [49] As Malarkey attempted to keep the conversation going, Brown told him he didn't have time to catch up. But the two would end up linking up in the future.[50]

Once the men were processed at W Company and accepted for duty in the 506th, they were marched over to the E Company area at Second Battalion. Period photographs of the camp show twelve rows of barracks buildings. Each company occupied a row of barracks, from Able to Item Company. Easy Company occupied one of the central rows of buildings, along one of the roads in the Regimental area. These roads were called, "company streets", and that was where Easy held its formations. The orderly room occupied the building closest to the main camp road. After that, the platoons were billeted in order from Third Platoon, Second Platoon, First Platoon, the company mess hall and then the company armory, in that order. The barracks buildings were constructed out of lumber, plywood and tarpaper and because of the flimsy construction, they were too cold in the winter and too hot in the summer.

Despite the poor construction, the camp's barracks provided the men a place to hang their uniforms and set up their sleeping area, which were a mix of single and double bunk beds.

There were morning inspections to ensure that bunks were properly made and if an inspecting officer or NCO could not bounce a quarter off the tightly made bed, the offending soldier lost his weekend pass or was forced to do some sort of remedial exercises or manual labor. Each man had a footlocker that sat at the end of his bunk along the aisle. In it, troopers placed their field uniforms, field gear, underwear, socks, hygienic paraphernalia, letter writing paper and pencils and any small mementos of home. Other uniforms were placed on wooden hangers and hung from a bar and shelf set that was nailed to the wall at approximately five feet high at the head of the bunk. This meager space was all that each man possessed.

These two photos provide an interesting comparison. On the left, Herman Hanson stands outside the Easy Company Headquarters building at Camp Toccoa. The company bulletin board is clearly visible as is the company sign. The photo on the right was taken at Camp Mackall and the company sign is now adorned with a parachute emblem. These photos symbolically show the evolution of Easy from recruits to paratroopers. (Left photo courtesy of the Hanson Family. Right photo, author's collection)

By the time late August rolled around, the company ranks had been fleshed out with 114 recruits and officers. As the men arrived at E Company, they reported to First Lieutenant Sobel and First Sergeant Morehead and were assigned to one of the company's three rifle platoons, or the company headquarters for those intended to receive specialized training in communications, administration or operational planning. Each platoon possessed two rifle squads, a mortar squad and a headquarters element. The numbers would change depending on a number of factors but generally each platoon had somewhere between 40 and 48 troopers present for duty.

There were eight officers, twenty-one NCOs, and one hundred and six enlisted men for a total of 135 troopers present for duty by the time Easy entered the training cycle in early September. Over time the numbers would fluctuate as men dropped from the training or were added. Attrition was high due to the intensity of the training and the overbearing methods of Sobel. Regardless, Easy Company would initially be defined by the shared suffering of Toccoa and that shared suffering began with the training regimen which was symbolized by the mountain that led to their regimental motto, "Currahee" or "Stands Alone."

2

"That Damn Mountain"

"The credit for the high attrition rate of men at Camp Toombs goes to Currahee Mountain"
– Richard Winters

While the training was rigorous and challenging, early on Currahee Mountain became the measuring stick for the paratroopers of the 506th Parachute Infantry Regiment, including those of Easy Company. Don Malarkey recalled, "At the barracks, I started talking to two guys who lived close to me in Astoria, Oregon. I kept questioning them about this damn mountain I heard so much about (while at W Company). One of them, Tom Burgess (Rod Bain was the other man), said to me, "Put your trunks on, we'll run you up there." So, the very first night I'm there, after they'd already run it once that day, these two guys decided to run me up there to see if I could do it. I didn't do it as well as they did, but I made it through. That was my baptism of fire in Toccoa."[51] That first run certainly took its toll as Don also recalled throwing up after returning to the barracks.[52]

Ed Tipper thought that they would not be made to run Currahee until the men had made it through the entire training program, as if it would be their "graduation exercise." On his first day Easy, someone blew a whistle and Ed, along with all the other members of the company, were ordered to change into their PT trunks, and they were off and running. So much for the "graduation exercise." This was discouraging to the aspiring paratrooper. "We ran most of three and a half miles to the top of Currahee and most of the way back down on a kind of a rough road that we all got to know very well. Few of us were really in shape to make that run, and some men dropped out of the unit that very first day."[53] Little did Ed know he and the others would be running up and down the mountain almost every day.

On top of all of their training, the men of Easy Company had to conquer Mount Currahee.
Left: The main trail was and is winding and ungraded, making for unsure footing. (Photo courtesy of the Currahee Military Museum via Brenda Carlan)
Middle: Once at the top, you are treated to a stunning view of the Georgia countryside as seen in this color photo taken in 1942. (Photo courtesy of Amos Taylor)
Right: "Bull" Randleman stands atop the rock formation near the pinnacle of Mount Currahee. (Photo courtesy of Denver "Bull" Randleman via Chris Langlois)

Winters described the pace and method for running Currahee with the troops. "The paratrooper run is a shuffle. To move a company of men up Currahee, you led them off at a paratrooper shuffle pace, then, as you felt the ranks falling apart due to stress, you cut the pace back to 'quick-time' march. After the ranks closed up again, and the troops were breathing normally, you went back to the double-time shuffle. The last mile up Currahee was done more at a quick-time than double-time pace. However, since most of the troops were so exhausted by that point they didn't know what they were doing. They thought quick-time was double-time...I can't remember anyone who could 'run' up Currahee."[54] Pat Christenson remembered the pacing much the same as Winters, although he added additional insights. "You would run until you thought your lungs would burst, praying for the command of quick-time to be given. When you were beginning to think that you could not go another ten feet more, you would hear the command, 'Quick-time...March!'"[55]

Bill Wingett captured the difficulties of dealing with Currahee in his vivid description of the mountain. "I remember a lot about Mount Currahee...At that time, it was just a road being used to get up there. Whoever went up there in a vehicle clawed their way up. There were ditches that ran back and forth across the road from erosion. There were rocks the size of baseballs, when you were trying to run, they would roll down on you. It was just a very treacherous place to be running either up or down." Paul "Hayseed" Rogers, who at 24 was considered to be an 'old man' by his fellow troopers, said, "The big thing he (Sobel) wanted was for us to be in tip-top shape — rough, tough." He remembered they ran three miles up a nearby mountain and back down every day as part of their training. Rogers, a "strong-as-an-ox" Kansan, ended up in Third Platoon.[56] [57]

These 1942-dated photos of the trail up Mount Currahee give the viewer a good picture of what it was like for the runner. (Photos courtesy of the Currahee Military Museum via Brenda Carlan)

Walter Gordon recalled the start of each training day vividly, and it revolved around running, and many times it meant dealing with Currahee Mountain. "We were issued blue trunks, not unlike tennis trunks that you'd see today...and every morning after reveille and chow, we'd put those trunks on, actually we had them on under our coveralls to have chow. We would shed the coveralls, fall out, and either we would run a stipulated number of miles or up this mountain. We had G.I . shoes at that time, because we had not been issued (paratrooper) boots...and there was never any stopping. Airborne troops never stopped. Your rest period consisted of quick time, which was marching. If they (the leadership) wanted to let you rest, you were reduced to quick time, but otherwise you were jogging...or trotting...is the way we moved from point A to point B."[58]

At one point, First Platoon man William "Shep" Howell tried to 'game the system.' Figuring he could creatively avoid the morning run up and down Currahee, he waited until the company got to an early bend in the trail. Seizing the moment, he sprinted out of formation and attempted to hide in some bushes. Unfortunately for the young Ohioan, he was spotted. Punishment from Sobel was swift. Howell was made to complete the run they were already making, and then when he arrived back at the bottom, he had to turn around and run it all over again. After that grueling episode, "Shep" never attempted to get out of a morning run again.[59]

Much of the course up the mountain featured a forty-five-degree uphill climb, which was reversed after reaching the summit. The initial acceptable mark for the individual run was 60-minute round trip. Easy also ran the mountain in company formation. Formation runs were more difficult because they were run at a slower pace, what Second Lieutenant Winters had previously referred to as the "Airborne Shuffle". Despite the difficulties, by all accounts this method was preferred as it allowed the men to support one another with words of encouragement or a shoulder to lean on. Camaraderie became the crutch that all the Easy men leaned upon when times were tough.

In an article for the inaugural issue of Para Dice Magazine in March 1943, Third Platoon man John Connell wrote about the difficulties of running Currahee and the camaraderie that it engendered. "I didn't realize then that it would become one of the strongest bonds ever to grip an outfit like this one. But I'm getting ahead of my story and I'll start from scratch, everyone couldn't write this story that is unless you have lived, trained, slept, and eaten with the 506th. I saw them run Currahee till they dropped and got up to run again. Currahee was the mountain in Toccoa; a 6-mile run, it was not fun. We swore at it from top to bottom, but we lived to run it again day after day. We saw and had sweat run into our eyes till it burned and kept plugging until we dropped. What kept us going, I told you I couldn't explain but everyone (sic) of the 506th will tell you it's there."[60] What the lanky Pennsylvanian was describing, the "it", was the camaraderie. They leaned on each other to survive the mountain.

It is easy to understand why "Shep" Howell thought it would be possible to pop out of formation and hide in the underbrush to avoid the daily run. Unfortunately for him, his little trick was discovered. It is amazing to think that Sobel didn't remove him from the company over that little stunt. (Photos courtesy of the Currahee Military Museum via Brenda Carlan)

Burr Smith reflected on that camaraderie with his close friends, a bond that only became stronger as the training got tougher.

"Skippy (Muck) was a happy guy, and those who knew him basked in the warmth of that happiness and were happy too. His closest friend, and, inevitably one of mine, was Don Malarkey, another warm, friendly and happy-go-lucky individual who likewise rose to the top of my list of personal heroes like cream to the top of the old-fashioned glass milk bottle."[61] Like Smith, Frank Perconte relied on his close friends too, saying that the only thing that made the training any easier was the presence of "good buddies like George Luz, "Gordy" Carson and Albert Blithe."[62] But for the most part, the platoons remained very insular. Lipton described it as such, "each platoon concentrated mainly on itself. We hardly knew the men in the other platoons in our own company and paid no attention to the other companies (and) of (the) men in them."[63]

While Sobel typically led the runs up and down Currahee, it was Clarence Hester's duty to bring up the rear. As such, he spent a lot of time collecting up the stragglers, especially in the early days of the company's training. "I brought up the rear, (and) carried a few football heroes. (They had) no stamina, no guts...(we) ended up with lean and mean 130 to 160 pound(ers). (We were) sorting (out the) too fat, too thin, no guts. We were the BEST and in that, our men were convinced they were the BEST!"[64] Platoon Leader Salve Matheson provided a bit more insight to how the Currahee runs were used as a tool to weed out non-hackers. "Officers were allowed to fall out once, before they were shipped (out), and enlisted men twice."[65]

Two of the more well-known photos show soldiers of G Company, 506th Parachute Infantry ascending (in coveralls) and descending (in PT gear) the trail. (Photos courtesy of James "Pee Wee" Martin via Doug Barber)

On a later Currahee run, Tipper received some unexpected and unspoken encouragement from one of his fellow troopers, as recounted in Marcus Brotherton's, *We Who Are Alive and Remain: Untold Stories from the Band of Brothers.* "He (Tipper) gasped with the other recruits, sides aching, legs pounding, wondering if he had the necessary stuff to make it into the paratroopers. Ahead of him ran one of the senior citizens of the bunch, Jack Ginn, already an old man at 25. Ginn had been out drinking the night before and stank, Tipper reported. As the sweet ugly smell wafted back, Tipper felt a strange sense of determination. Ginn continued to run, hangover and all, and Tipper figured that if Ginn could make it to the top, boozy as he was, then certainly he would be able to make it, too."[66]

Rod Strohl and Forrest Guth decided if they had to train at Toccoa all over again, one thing they would have eliminated was the runs up and down Currahee Mountain. "I think it was the worst thing," Guth stated flatly. Strohl echoed the sentiment. "I would have cut it out (running the mountain)."[67] Regardless how much the men hated it, Currahee became the symbol of the 506th Regiment, and it symbolized the exacting training that they had endured. While Currahee created lasting memories for the men of Easy Company, nothing left a more lasting imprint on them than Herbert Maxwell Sobel.

3

Sobel Leaves His Indelible Mark

"Each man in this company will learn the importance of discipline and practice it or he won't remain in this unit for long. If you don't complete your assignments or pass inspections, you'll receive company punishment. If you continue to fail to accomplish what I consider your duty, you'll be disqualified from the Parachute Infantry."

— Pat Christenson quoting Herbert Sobel's introductory speech to the men of Easy Company.

The men began to realize what they had gotten themselves into when training began in earnest. It was tough...tougher than anything they had experienced in their young lives. Even then, the training was made all the tougher due to the exceedingly strict training methods of First Lieutenant Sobel. Walter Gordon summarized it as such; "E Company, we had this natural instinct to (bond). The blending of the enlisted men with their officers who had been assigned to them, all of that took place (at Toccoa). But E Company was unique because we had another element which featured prominently into our later being such a fine and close unit. And that was the nefarious Herbert M. Sobel."[68]

No one would have a more lasting effect on Easy Company than Herbert Sobel. He made it his personal mission to develop Easy into a well-trained parachute infantry company, and he never avoided his duty.

Left: Sobel, seated third from the left, awaits take off with the rest of the jumpers, for one of his qualifying jumps at Toccoa. (Photo courtesy of 506th Infantry Association Website)

Right: Sobel, with 1903 Springfield rifle slung on his back, negotiates a wall obstacle. As Don Malarkey told me, "He never told us to do anything he wouldn't do as well." (Photo courtesy of Michael Sobel)

Despite having the even-tempered First Lieutenant Clarence Hester as his executive officer or "XO", and even keeled and conscientious platoon leaders like Winters, Salve Matheson and Wally Moore, Sobel strove to find the smallest infractions.

He was under orders from Sink and Major Strayer to weed out the weak so that only the cream of the crop remained. He took that guidance to the extreme. Of the first 114 officers and enlisted men present for duty with E Company on 25 August 1942, only 35 would go into combat with the unit on D-Day, 6 June 1944.

While the enlisted men suffered through the training day-to-day, at the end of the day, they could retreat to their barracks or head over to the Post Exchange, or PX, to listen to the jukebox and slug down soda pops and candy bars. For the officers, it was like fighting against an exceedingly strong current with no relief. Eventually most of them would be overtaken and in time, accepted a position on battalion staff to get away from their tyrannical company commander. In short, Sobel leaned on the officers even more heavily than he did the enlisted men. "Rook" Rader observed that Sobel was, "the same or maybe more severe with his company officers as he ever was with his non-coms and enlisted men."[69]

For the enlisted ranks, grousing about Sobel and his methods became more vocal in October 1942. Complaints began in earnest on the 9th of October when the men undertook their weekly Friday night march. After returning to the barracks after midnight, they were awoken at 0700 hours. Still stiff and achy from the march, the men stood at attention while Sobel barked at them. "We are running Currahee...now!" And off they went. After they returned, the men quickly showered and prepared for the normal Saturday morning inspection. Even though they were sore, spirits were high in anticipation of getting out of the camp and blowing off steam in town. Unfortunately, the inspection did not go well. Sobel, displeased, ordered them back into their PT gear. "You men do not deserve a pass! Your barracks are unsatisfactory! Since you can't maintain your weapons, equipment and uniforms to the Army standard, we'll run Currahee again!" And off they went again...grumbling the whole way.

The enlisted men, many of whom had come directly from civilian life, had no measuring stick to frame Sobel's methods. Most just figured it was the "Army way." Those that had spent time in the notoriously easy-going National Guard figured this was what life in the real Army was like. Leaders above the company level could see what was going on, but they had their orders to weed out the weak. Sobel was doing just that. "Sobel was pretty hard-boiled, and the men (recruited) just out of civilian life, they didn't cotton to that," remembered Battalion Commander Robert Strayer. "He just didn't have the finesse to get across as a leader to the men."[70] That point would come to a head in the future, but for now, Sobel did his job and did so to the extreme.

On the 21st of October, the company was scheduled to run half of the course up the mountain, with a turnaround that would make the course three miles in total. In mid-run, Sobel decided to make the company run the whole six miles, up and down. It was pure mental warfare. To add insult to injury, Sobel removed the recovery time between the Currahee run and the next training event, log rolling and throwing between squads. Second Lieutenant Winters thought, "It (the training) was on the schedule, but not to run to the top of the mountain and try and kill the men."[71] Most of the men were so new to the service that they had no idea what the difference was between normal and excessive levels of training and discipline. But by late October, they could easily see that no other company in the battalion was being held to the same standard as they were.

Richard Winters was a leader that everyone respected. A gifted athlete, he also managed to walk the fine line of being approachable yet maintaining objective separation from his men. This allowed him to be focused on helping them improve as soldiers, while permitting him to make difficult decisions without emotional baggage having been formed.
(Photos courtesy of the Richard Winters Collection, AHEC)

Winters, an officer all the men looked up to and respected, "Gave Sobel absolutely no credit for making Easy better than it would have been under someone else...Sobel did set a hard training regimen, but...he basically followed what Battalion wanted him to do. The only thing he did differently was add the chickenshit and bust on guys for minor or imagined infractions. Worse, Winters said, when training moved from the physical (calisthenics and running) to field work, map reading and leadership skills, Sobel was sorely lacking and compensated by - as Winters said – (by) "just getting mean."[72] "Rook" Rader effectively captured the essence of chickenshit by providing some examples. Sobel would, "gig your ass for not shaving (despite most of us only had fuzz on our cheeks) or (saying you had) lint in your (rifle) stacking swivel."[73] Gigging the men for such minor infractions and punishing them to the fullest extent possible for them, fostered a high level of animosity for Sobel among the enlisted.

Not only did he alienate the enlisted men, but he continued to alienate his own officers as well. Talented platoon leaders like Lewis Nixon, Matheson, George Lavenson, Archibald Barnwell, Wally Moore and James Davis either left the paratroops, or took well-earned positions on the Second Battalion or Regimental staffs. While Nixon, Matheson, and Lavenson moved up to the Regimental staff, Moore would end up joining the fledgling Pathfinders in England. Walter Gordon considered Moore the cream of the officer crop; "...(he was a) fine, fine young man. (He) was small of stature but a giant of a man. There was no familiarity to his makeup, but he impressed us as an extremely capable man and knowledgeable and we all looked up to him and he did a great deal in molding the third platoon of E Company into a very cohesive and fine group."[74] The loss of Moore would be felt, but that would take place in the future. Despite the grousing about how he was accomplishing the feat, Sobel's methods undeniably shaped one hell of a capable rifle company. The men would finish out their training in amazing physical condition (because they were constantly being punished with additional PT), and bonded due to their common enemy, Sobel.

Three of Easy Company's early platoon leaders.
Left: Lewis Nixon, began his career in the 506th as a platoon leader in Easy Company. Talented and brave, Nix would serve in a number of S-2 (Intelligence) and S-3 (Operations) billets across the 506th during World War II. (Photo courtesy of the Richard Winters via Erik Jendresen)
Middle: Wally Moore, as Walter Gordon stated, was a "fine and conscientious young man," and while "short in stature, was long in common sense." He led Easy's Third Platoon. (Author's Collection)
Right: Salve "Mat" Matheson was an outstanding platoon leader who excelled in teaching his young soldiers. Colonel Sink saw immense potential in Matheson and brought him to Regimental Headquarters. (Author's Collection)

Yet, despite the already difficult training, Sobel continued to make things worse. He constantly focused on the unimportant...the 'chickenshit' as the men were fond of saying. While attention to detail was vital to almost every task the men were confronted with, Sobel increased the pressure by nitpicking the men incessantly. He constantly gigged the men for minor infractions. Such was the case with Frederick Bealke, a private in Easy's Second Platoon. Sobel consistently gigged Bealke for not shaving despite the private's inability to grow even a whisker. Don Malarkey said that Bealke probably never owned a razor, and with good reason. He didn't need one. He couldn't grow much more than peach fuzz, which was only visible in the sunlight. Sobel would lean in close to Bealke's face, scrutinizing the young man's cheeks. As soon as he spotted the tiniest of hairs, Sobel would rescind the entire platoon's passes for the infraction. Finally, the men in Second Platoon, frustrated with the lack of passes being handed out, held Bealke down and shaved him.[75]

"Skinny" Sisk was a frequent visitor of Sobel's doghouse. The Third Platoon man was notorious for getting into trouble. A late-night rendezvous on a train track with a young lady and an oncoming locomotive earned "Skinny" a several days stay in the Toccoa sheriff's jail. About the incident, he often stated with a laugh, "I was coming, she was coming, and the train was coming...and none of us were stopping."[76] Even the modest Winters had to chuckle about the incident as he was springing Sisk from the local lockup and transporting him to the Regimental guardhouse.

In addition to his time in the hoosegow, "Skinny" often found himself the recipient of another favorite punishment of Sobel's; the digging of a six-by-six-by-six pit. Offenders were required to dig a 6-foot wide by 6-foot long by 6-foot-deep pit. Once an officer inspected it (it was often Sobel), the trooper was made to fill it right back up. At one point, Sobel called Sisk out to the front of the company and crowned him, "The King of the Foxholes," for how many he had dug due to his frequent rule breaking. James "Moe" Alley of Third Platoon also recalled being made to dig a 6x6x6 pit on several occasions. Still, occasionally, the men got over on their company commander by beating the punishment.

"Skinny" recalled a particular instance when he was assigned to dig several foxholes as punishment. Sobel detailed Second Lieutenant George Lavenson to march Sisk out to the training area and watch him as he dug. As "Skinny" told the story, he had hidden out in town to get some beer while the rest of the company was out on a field problem and Sobel ended up catching him. "When we came back he assigned me a duty to walk back to where we started the problem...I was to dig ten foxholes. He sent a lieutenant with me, George Lavenson, I believe, to make sure that I would dig the foxhole. In the meantime, I loaded my backpack with beer. We proceeded to march out about six or eight miles; I forget how much. Anyway, I had dug down far enough to get my feet down in the first foxhole and sit down and open me up a bottle of beer, and started drinking it. Then I scratched down a little more and opened up another one. Lieutenant Lavenson asked me if I was going to sit there and drink all that beer and let him die of thirst. I told him, 'Noooooo, you can have some of this if you want to dig some foxholes.' So, naturally, he didn't care too much about that. I just kept on drinking, kept on digging, not too hard. But anyway, he decided he wanted one of them damn beers, that's what he said.Let's just stay out here a while and drink this beer and we will go back, and I'll just say you dug them and don't you ever tell on me. After he brought me back he called the company out...and crowned me "king of the foxholes."[77] Lavenson was a good as his word and Sobel never found out about what the two had done.

Left: George Lavenson started out as a platoon leader in Easy and eventually transferred to 2nd Battalion Headquarters as the S-1 (Personnel) officer. He oversaw Skinny Sisk's foxhole digging punishment. (Author's Collection)
Middle: Skinny Sisk managed to bribe Lavenson with beers in order to get out of digging. (Photo courtesy of Robert Burr Smith via Susan Smith Finn)
Right: Joe Toye quickly tired of Sobel's overbearing methods but chose not to make waves in order to remain with Easy Company. (Photo courtesy of the Toye Family)

Unfortunately, the men didn't get over anywhere near enough to alleviate the stress that was building in the company. Men like Ed Tipper and "Smokey" Gordon built up a hatred for Sobel that they carried throughout their lives. They even devised a nickname for him...the Black Swan. They felt that Sobel, with his shock of black hair, gangly arms and legs, and odd running style, looked like the waterfowl of the same name. And within the ranks, muttered under many a breath, the name stuck. Joe Toye accurately captured the men's collective feelings about Sobel. "I didn't think much of him. He tried to do too much. I don't know where he got this business of being so strict. He used to pull things that no other officer ever pulled, and I'd seen that everybody maybe Saturday afternoon would be going to town, and he'd pull us out for some kind of inspection or some little thing that, you know — I don't know. (He) just didn't fit (in)."[78]

There was always some idiosyncrasy on Sobel's part that forced the men to question his sanity and his future in the paratroops. Johnny Martin thought his actions were questionable at best, but he also surmised that there was something else motivating the lanky officer. "Jesus, what kind of an outfit is this?!" Martin said. We'd get out there to run and he'd say, "Hi Ho Silver!" and all this stuff. After a while you think, "Come on, let's get off of that stuff. But anyway, I think Sobel ran each course simply because I don't think the rest of the officers including the Colonel and all of that felt that that was the best job for him." If that was how Sobel felt, he never let anyone know but he had also obviously decided that he was going to attempt to train the most proficient and disciplined company in the regiment. To sum up his view of Sobel's disciplinary methods, Martin provided some thoughtful insight into what he believed to be Sobel's motivations. "I think probably Sobel got the (rule) book out more than a lot of other company commanders and the rest of the battalion might have done. I think he was a little more of a task master, you might say. Sobel was a tougher company commander with respect to living up to the rules and doing the things that was (sic) necessary to build a good company. And I think the guy had his heart and soul in it. There's no doubt about it."[79]

Robert Rader concurred with Martin's assessment of Sobel. "I do believe this, I think he sacrificed his character, personality, mind, and body to weld together men (boys – average age of 19 years) who met at Toccoa, Georgia, that hot August day." Rader also knew that, with an ongoing global conflict, that E Company needed to be ready to do its duty, and he understood that Sobel was doing his best to prepare them for that eventuality. "He (Sobel) appeared to take on the responsibility on his shoulders the problems of the whole warring Free World. He did the best he could, and by that token, he expected every man to...know your weapons, equipment, and your strengths, both mentally and physically and take care of all things that are military until this crazy conflict is over. Once the company was formed for training, he urged, cajoled, drove, trained, ordered, 'ate' anybody out for the littlest of things, (a) real 'chickenshit' attitude. (But) he made sure the company was fed, clothed, and trained and he didn't care on the surface whose feelings got trampled on."[80] [81]

While Robert Rader (left) understood the method in the madness of Captain Sobel, Johnny Martin (middle) soon grew tired of the "chickenshit." George Luz (right) never missed an opportunity to have a little fun at his company commander's expense. (Photos courtesy of the Carson Family)

Some of the other men also appreciated Sobel's methods, as they understood what he was trying to accomplish. He wanted the men to be the best at everything. "Buck" Taylor said, "We did a lot of fieldwork and a lot of physical exercise training. Sobel was a good training officer. He was strict and he pushed the fellas. He went right with them. We went up Currahee in the morning on our runs and usually he was right there with us. He was great and whipped us into shape and I feel that most all of the fellas today would say he was responsible, really, for our good physical condition."[82] But Taylor expanded on his description of Sobel and it brought to light the extent of the hatred towards their taskmaster company commander. "He was a hard driver, and a lot of the fellows disliked him for this reason, me included. But looking back on it, he really did make the company. He shaped us up, he made us what we were, but...everybody was quite concerned — especially the NCO's, and I'm sure some of the officers too — quite concerned about him leading us into combat. He was a little bit too hyper, high-strung, and everybody was afraid that he would make the wrong decision that would cost some lives, and there were several fellows in the company who were threatening to get him if they ever got into combat, they disliked him that much."[83]

One thing that many of the men noticed was that Sobel never ordered them to do anything he would not do himself. Sometimes this worked against him. While running Currahee Mountain, George Luz would often impersonate one of the senior officers, calling out, "Captain Sobel to the rear!" Since Sobel always ran at the head of the formation, he would double back to the rear, only to find out that there was no senior officer present. He would then double time back to the front of the company, burning up precious energy while eliciting muffled laughter from his amused troopers who were well aware of Luz's mimicking ability.[84] Still he never quit. He may have struggled at times to complete some of the physical activities, but he made it through, nonetheless.

Robert Rader saw Sobel as, "...a man possessed with the desire to make top-notch troopers out of cowboys, mule skinners, students, soda jerks, dancers, and many other occupations too numerous to mention." Despite this insight, it didn't make dealing with the 'chickenshit' any easier. According to "Shifty" Powers, "He was real, real strict, but he never asked you to do anything he wouldn't do. We would run 6 miles up that hill and back down. Whatever we did, he did. He was strict as he could be. He kept you in camp on Friday nights, but he would make sure you were entertained so you wouldn't be out running around getting into trouble. He was a good training officer."[85] Dewitt Lowrey echoed Powers' sentiment, "Sobel ...did what he had to do to keep the boys in line."[86]

"Popeye" Wynn had this to say about Sobel, "He pushed everybody hard. He stayed with us and took all the physical abuse we did. He didn't want 'E' to be second best. He was very hard on me, but I goofed off a lot and when he caught me, I paid."[87] "Smokey" Gordon agreed with "Popeye's" assessment of Sobel's objective in saying that, "In spite of the fact I was still in shock, which resulted from being thrust into this hostile and strange environment, I felt that our commander was actually giving his best effort towards making us the best in the regiment."[88]

Despite his loudly professed dislike for him, Malarkey too noticed that Sobel took part in the training. "For all the criticism about Sobel—and I participated in a lot of it I'm sure—whatever we did, he did. If we were on a forced march at night, he would be there. If we had to run the mountain, he would run the mountain. He wasn't just a dictator; he would do the stuff that we had to do along with us."[89] More of the men understood this than would care to admit...even with the pain, the heat, the suffering and the mundane repetition of events, they understood that Sobel was molding them into a unit. In the immortal words of Rod Strohl, "Sobel MADE Easy Company."[90]

Despite whatever understanding they may have had at the time, the men still felt targeted. They couldn't even escape Sobel's chickenshit ways while on field problems, where combat tactics kept the garrison type activities to a minimum. Garrison was where Sobel controlled every aspect of the company's lives. In the field, the men were solidly under control of their platoon leaders, but even then, Sobel found ways to nitpick them. Like many of his buddies, Joe Toye experienced Sobel's wrath but for something he felt was unjustified. "Somebody took me on an outpost, with Warren Muck, (who was) a good friend of mine. But this (field) problem was still going on. One of us was supposed to stay awake. Somehow or other, two of us fell asleep, and the officers came around and he caught us. So, the next day, Sobel called us down to the office, and he said to the both of us, he said, "You can have your choice. You can either have a court martial or company punishment."...What the hell am I going to take a court martial for? I said, "Oh, I'll take company punishment." It started with this company punishment. First, the next day, we had to go down, clean all the stoves in the thing (all of the company barracks buildings). Then we had to march behind trucks out to this area where all these guys were throwing their C (ration) cans and clean all that up along the road, and come back."[91] Since they were close to heading to Benning for jump training, Toye, like many of the other men, took the punishment to avoid being thrown out of the company. Their desire to become paratroopers overrode their hatred of Sobel's methods. "Popeye" Wynn said it best, "I feared being washed out!" and because of that attitude, many of the men accepted the seemingly endless stream of punishment.

4

Sink's Muscle College

When they arrived at Toccoa, "It was nothing but obstacle courses, weapons training, push-ups, running up a mountain named 'Currahee,' which became our battle cry," said Johnny Martin. "He (Colonel Sink) wanted to weed out the weak ones."[92] And weed them out he did. Ed Tipper also recalled the difficulty of the training. "The paratroop training was far more difficult than I expected. We were all tested to our limits and beyond daily, and every few days men were eliminated and sent out to the regular infantry."[93] Don Malarkey remembered the hard training well. "Most of our training at Toccoa was primarily physical in nature, and very little in actual military training. We did a lot of running in the mountains, and when we weren't doing that, we were busy on obstacle courses and calisthenics, besides working on mock-up towers that were to be used for our training in the jump school when we went to Ft. Benning."[94] Each training day was a whirlwind of activity, from daylight until dark. Troopers came and went quickly. Some were injured. Others could not handle the physical pace of the training. Many of the men recalled going to sleep and waking up the next day with an empty bunk next to them.

The obstacle course was a key component in the training program to get the men in shape and harden their bodies, and they ran it often. Carwood Lipton described the course as such, "(It was) diabolically laid out...wound around a steep slope with cut-off trees, crotch-high, to emasculate the runner if his foot skipped; walls to bar-vault; ropes to traverse hand-over-hand; logs over water to run across; a rope to use to swing over a ravine; and an uphill finish that drained the last of our stamina. Each man was timed every time he ran it, and the was no room in the company for those who could not meet the required time."[95]

The obstacle course was not for the old or faint-hearted. It required speed, agility, mental and physical toughness. And it was rugged, with rough-hewn logs and steep up/downhill gradients. (Top left photo courtesy of Raymond Scully via Chase Blackwell, top right photo courtesy of the Carson Family; bottom row photos courtesy of James "Pee Wee" Martin via Doug Barber, The Currahee Scrapbook)

Walter Gordon concurred with Lipton's description, recalling that they ran it every other day, and described it in a very tongue-in-cheek manner, "(It) was kind of in a depressed section of the camp enclosure. And they had a series of various contrivances designed to kill you, of course, and that consisted of crawling over vertical walls, or climbing up ropes with knots in them, walking across a log, across a stream or wet pit, going up a hill, negotiating all sorts of obstacles, hence, it was called the obstacle course."[96] Lipton and Gordon were not far off in their assessment of the course's difficulty. Private John

Connell witnessed one trooper fall off the ladder bars breaking an ankle and another injuring his back while negotiating another obstacle. The course was not for the physically unfit or the faint of heart.

These photos permit the view to observe the level of physical strength and agility required to negotiate the obstacles. In the first, Dick Winters and another officer climb the log wall. The second and third pictures depict members of G Company gingerly walking across the log and bar sets over the water obstacle. (Left photo courtesy of the Author's Collection, middle and right photos courtesy of Raymond Scully via Chase Blackwell)

There was also another type of physical activity that stuck out in Gordon's memory, "One of the more strenuous things was a pine log cut in probably 25 foot lengths, where a group of us would pick up this log and have to raise it, lower it, pitch it in the air, catch it, whatever they could think of." Lipton expanded on the physical training regimen, "Push-ups, chin-ups, dashes, runs, and every conceivable exercise to develop our legs (to stand the shock of parachute landings with an extra hundred pounds of equipment) and our arms (to control our chutes during descent and to give the downward pull just before landing) were daily fare. We paired off and wrestled without reservation or pity."[97] [98] As Lipton observed, many of these exercises were designed to build their muscles up to withstand the rigors of Airborne duty. Of course, the men, as raw recruits, did not know this yet.

Left: Walter Gordon recalled many sessions of "Log PT" the men of Easy took part in while at Camp Toccoa. The goal of exercises like this was to build up upper body strength and stamina. These paratrooper trainees at Fort Benning are taking part in the log toss. (Author's Collection via the U.S. Army Signal Corps)
Right: A Currahee trooper, with a firm grip on his "shroud lines" descends from the 34-foot tower at Camp Toccoa. "Log PT" helped build up the muscle groups that were vital for manipulating the shroud lines with descending a parachute. (Photo courtesy of the Atlanta Historical Center via Sue Hardy VerHoef)

In addition to the calisthenics, the men were fed a steady diet of infantry tactics, map and compass instruction, weapons training and employment, hand-to-hand combat, which included wrestling, and bayonet drill. Winters had been an excellent wrestler in school and frequently challenged the other men in matches during training. With the strength developed from farming, Third Platoon's Paul Rogers was a formidable wrestler as well. "I could throw anybody in the platoon, wrestling, out (of the ring). You know, I was strong." The pair squared off one time and for once, Rogers found an opponent that could match his prowess. "I wrestled with him (Winters). He was a good wrestler. He showed me a little about it."[99] James "Red" Benton, from Easy's First Platoon, had been a circus acrobat before joining the army. He also gave Winters a run for his money in the wrestling ring.

The men went through hours of close order drill to learn to march in formation and more importantly, learn instantaneous obedience to orders. Gordon, who by admission, wasn't a very good soldier at Toccoa, said of receiving orders, "The one thing we had learned, and one thing that had been drilled into us was that if you received an order, you had best carry it out as well and fast as you can." For example, upon finishing runs, Sobel would not release them until each and every man was standing ramrod straight and still. He would run up and down the ranks of the company yelling, "Stand still! STAND STILL!" Only when they met his standard would he release them.

If the men carried or used their weapons during the day, they were required to clean them and have them inspected by the noncoms before they were permitted to go evening chow. Pat Christenson remembered, "After carrying your weapons a good part of the day you spent at least an hour of your own time cleaning them before chow time, if you were fast enough."[100] Taking your time cleaning could and sometimes did mean that you'd miss your evening meal. In the event of rainy weather, the men could and did spend additional time cleaning weapons, along with taking classes on map and compass reading. According to Winters, "Every minute of time was spent honing the squads, (platoons), and companies into a fighting unit."[101]

Even chow became something to endure instead of something to enjoy, as it oftentimes consisted of "SOS" or "shit on a shingle," in other words, chipped beef in gravy on toast. In fact, the food was usually so bad that Christenson said that the men would often argue with the cooks over its quality, or lack thereof, in the chow line. Sometimes, instead of criticizing the food, the men took out their frustrations on the cooks through verbal abuse. But you had to be careful doing so because frequently company punishment would come in the form of K.P. (Kitchen Police) duty, and since most of the cooks were noncoms, they would exact their revenge on any private that had berated them too loudly. Barring any incidents, the men would line up with mess kits in hand, go through the line, get their food, and then find a place to sit down and eat. If the mess hall was full, they would go outside, find a spot against a tree or building, and wolf down their food.

Meals were something to endure, not enjoy.
Left top: Members of Easy's First Platoon file past the field kitchen tent to fill their mess kits. (Photo courtesy of the Carson Family)
Left bottom: Several members of First Platoon relax in the field while eating their meal. (Photo courtesy of the Carson Family)
Right: With canteens and canteen cups in hand, three First Platoon troopers enjoy some toast and a cigarette, while taking a break on a conditioning hike. (Author's Collection)

One of the few memorable times that they did have decent chow was the day they were treated to a "tremendous meal of spaghetti" for lunch, but according to Ed Tipper, they were soon ordered to run Currahee Mountain shortly after eating. He vividly remembered his fellow troopers vomiting up spaghetti all along the trail, and there was Sobel exhorting them to quit and head down the mountain to the ambulance, and an eventual discharge from the paratroops. According to "Tip," "those who dropped out of the run and accepted the medics' invitation to ride back in the ambulances found themselves shipped out that same day."[102]

One weekly event that stood out in the minds of many of the men was the Friday night marches. When the company first started doing them, they were between 7 and 10 miles in distance. Gradually Sobel added miles to them. While many of the other companies were being released on passes for the weekend, Easy was out trudging through the Georgia countryside with weapons and full equipment. Since company duties would generally keep Sobel occupied on Friday nights, he would give the ranking officer (the duty rotated among the platoon leaders) a compass and objective, along with a time limit in which to finish the march. The company would step off at a fast clip, and it was not unusual for them to march for several hours without breaks. If someone needed to answer the call of nature, they did so while the company continued to march. Then the trooper would have to double time to catch back up with the formation.

Of the young troopers, it was the machine gunners and mortar men that had it worst due to their heavier loads. Walter Gordon devised an ingenious way to lighten his load on the marches by altering his M1919 .30 caliber machine gun. Gordon realized that the barrel diameter was similar to that of a broomstick handle. So, he acquired a broomstick and cut it down to the same length of the barrel and painted it black. He then removed the jacket and real barrel and replaced it with the broomstick handle. He then slid the jacket back over the barrel, thus lightening his load by several pounds...that is until Sobel inspected his machine gun after one march. Punishment was swift...loss of a weekend pass and the digging of a 6x6x6 pit. Sheepishly, Gordon complied.[103] What else could he do if he wanted to remain with the company?

These photos provide a rare glimpse into the early days of Easy Company. The tarpaper barracks are clearly seen, as are the HBT coveralls typically worn during training. The unit used the blackboard to explain basic tactics to its new soldiers.
Left: Walter Gordon and Ledley Pace man their M1919 machine gun for a photo after returning from a field problem. (Photo courtesy of Amos Taylor)
Right: Amos "Buck" Taylor and Ledley Pace pose for a photo. (Photo courtesy of Amos Taylor)

In addition to the heavy loads, the men were told to adhere to strict noise and water discipline or face punishment. The men marched regardless of weather conditions. In the muggy Georgia humidity, the men quickly saturated their issue coveralls with sweat. Pity the trooper that wasn't in his correct spot in the ranks when they returned to the camp. Sobel would be waiting to pounce on any infraction. He would even go so far as to check the contents of every man's canteen. According to Pat Christenson, "We felt miserable, and were always thirsty on the long marches with no water."[104] Many of the men reflected on the marches in that they instilled confidence in themselves that they could endure a certain amount of physical and emotional hardship.

The first Friday night march took place on September 11th, 1942, and was 7 miles in distance. Private Robert T. Smith, a member of First Platoon's Mortar Squad remembered that first march as it came as quite a shock. "As we were enjoying a movie we got called out and were marched seven miles. We didn't get in until one in the morning and so you can understand how we felt when they got us up at 5:45 Saturday morning."[105] Guth was more pragmatic in his outlook regarding the harsh training methods. He recalled the frequent hikes, many as long as 25 miles. "And we didn't get paid by the mile," he joked.[106] For "Shifty" Powers, the only thing to look forward to when it came to the marches was the good coffee and fresh doughnuts that were made available to the men after returning.[107] Although it didn't happen right away, after the men had been at Toccoa for a while, they were permitted to sleep until 7:30 or 8:00AM the next day. Of course, they had to form for parade soon after, but once that was completed, they were typically given the rest of the weekend off...if Sobel hadn't taken away their passes.

Friday Night Fights
First Platoon's Pat Christenson squares off against Second Platoon's Tom Burgess. A weekend pass is most likely on the line. Sobel, Joe Toye, Joe Liebgott, Dan West, Bill Guarnere, Moe Alley, and Don Hoobler are some of the troopers gathered around the pair.
(Photos courtesy of the Carson Family and the author's collection)

Another event that frequently took place on Friday was company boxing matches. Often, Sobel would only award passes to the winners of the fights. He also enjoyed pitting one platoon against another, feeling that it would build a competitive spirit within the company. While the men were friends, platoon loyalty at times trumped that of the company, because the men in the platoon became very close under the harsh conditions. One well-known sequence of photos taken by Burr Smith depicted Pat Christenson of First Platoon, taking on Tom Burgess of Second Platoon. Other men in the company, like Dan West, Joe Liebgott, "Moe" Alley, "Skip" Muck, Don Hoobler, Bill Guarnere, Joe Toye, and even Sobel, were gathered around the impromptu ring in the center of the company street, cheering on the combatants. Both men were swinging vigorously, and despite the use of oversized, heavily padded gloves, they were obviously trying to win. It was apparent that neither one wanted to lose out on a chance for a weekend pass, and with it, a break from Sobel.

Even though they were technically off duty, those troopers that remained in camp on Saturday mornings were subject to impromptu inspections prior to the weekly Regimental parade. Sobel would come through each barracks building and pour over the men's weapons, equipment and clothing. On some occasions, even if none were present, he would invent infractions. In his mind, he believed that this method instilled an appreciation for attention to detail in his troops. For those troops that did not meet his standard during those inspections, a weekend pass quickly went out the proverbial window.

* * *

The Regiment eventually arranged with Clemson University in South Carolina, for the troops to qualify with their weapons on the campus' Reserve Officers' Training Corps (ROTC) rifle range for one week in late September. Popular and highly capable Third Platoon Leader, Wally Moore, took a small detachment of men to Clemson to prepare the range and billeting areas. Staff Sergeant Steven Kudla, along with privates Joe Stickley, "Buck" Taylor, Ken Baldwin, Frank Perconte, Albert Mahmood, "Popeye" Wynn, Harry Stanfield, Carwood Lipton, "Red" Wright, Lavon Reese and Floyd Talbert did their best to clear the fields and ensure the range's targets were in working order. The remainder of the company, back at

Toccoa, gathered their equipment and marched the 32 miles to the school's campus.

Scenes from the Rifle Range at Clemson University
Left: After arriving, George Luz and John Sheely enjoy a moment of camaraderie. (Photo courtesy of the Carson Family)
Middle: An unidentified Easy Company officer mans the radio in the rifle "butts." This was where targets were lifted for the men to shot on. Once the rounds were fired, the targets were dropped back into the butts for marking and scoring. (Photo courtesy of the Carson Family)
Right: While the men slept in pup tents (scene in the left picture), the officers were billeted in squad tent, as seen here). (Photo courtesy of the Richard Winters Collection, AHEC)

The accommodations were on par with W Company at Toccoa, which is to say, conditions were extremely rustic. The latrine was nothing more than a couple of sturdy logs lashed together with shelter half tents providing the only privacy for the user.

The men drank water collected in a 36-gallon lyster bag that hug suspended from three tent poles, tied together in pyramid form. The men slept in open-ended pup tents or if they were lucky, squad tents which were erected near the range. Temperatures dropped at night, and the men were provided only a couple of wool blankets to help ward off the chill.

Despite the conditions, the men focused on shooting Expert. Paper targets were arranged along a berm at the front end of the range area. The men were placed in relays and assigned a target. Each was provided three rounds to 'zero' his rifle.[108] Once that was completed, the men practiced for several days before firing for an official score. Second Battalion medic Al Mampre was positioned near "Shifty" Powers on the range. At one point, he bet "Shifty" a candy bar that he couldn't hit the bullseye, "Eight candy bars later" they were done with that bet because Al was out of treats.[109] [110] "Buck" Taylor shot the highest score in the company at the range and earned himself a three-day pass, which he used to go see his girlfriend, Elaine.[111]

"Good Scores and Lots of Candy Bars"
Left: Shifty Powers made the most of his shooting prowess. (Photo courtesy of the Rhinehart Family via Jay Hopkins
Middle: Al Mampre learned a lesson about betting against Shifty in a shooting contest. (Photo courtesy of Al Mampre)
Right: Buck Taylor was one of the company's high shooters and earned himself a three-day pass. (Photo courtesy of Amos Taylor)

Things went less well for Carwood Lipton. A natural southpaw (aka left-handed) shooter, Lip decided to train himself to shoot right-handed while at Toccoa. He performed everything right-handed during all the rifle related exercises and dry runs. When they arrived at Clemson, Lip fired right-handed for qualification. It didn't work out the way he planned. "I qualified as only a Marksman, the lowest level, above it being Sharpshooter and then Expert. I changed back to left-handed after that and fired Expert on other ranges with both rifle and Tommy gun."[112]

While the riflemen were shooting at Clemson, the machine

gunners and mortarmen stayed at Toccoa to qualify on their weapons. Second Lieutenant Sal Matheson of First Platoon was particularly adept at training his young charges on the intricacies of M1919 .30 caliber machine gun. Because of this skill, Sobel charged Matheson with the training of all of Easy Company's machine gunners. This not to say that there was any method to whom was chosen to man the machine guns and other heavy weapons, like the 60mm mortar. Bill Wingett reflected on his induction into the world of machine guns and machine gun teams. "Being a machine gunner wasn't something I picked. I had picked being in the Airborne, but I certainly didn't get to choose what I was going to do or who was going to be my partner. If (Everett) Gray and I had arrived a day apart, we would have never been partners on the machine gun."[113] [114] The training was thorough and intense. The men were taught the intricacies of using the traverse and elevation knobs to precisely control the movement of the gun. They were also taught to regulate their rate of fire so they would not burn out the barrel and in case they did, they were also instructed on how to quickly swap out barrels as well. Under the excellent tutelage of trainers like Matheson, the men became highly proficient at laying down fire and maintaining their weapons.

Left: A First Platoon machine gun team made up of Woodrow Robbins (on the gun), Pat Christenson (assistant) and Shep Howell (to the rear) has set up their gun in a copse of pine trees. (Photo courtesy of the Carson Family)
Middle: The machine gun team of Everett Gray, William Dukeman, and Bill Wingett clown around with their .30-caliber while packing aerial delivery containers. (Photo courtesy of Bill Wingett)
Right: Machine gunner George Rajner takes a break during a field Problem. (Photo courtesy of Robert Burr Smith via Susan Smith Finn)

This is not to say the machine gunners had it easy. The 32-pound M1919 machine gun was difficult to carry and uncomfortable to say the least. "Smokey" Gordon recalled when he was assigned to be a machine gunner, "Contrary to what you might think, a machine gun is quite heavy and certainly awkward to balance on a shoulder. Pads were provided but I now realize I should have been furnished a frontal lobotomy as well. The only justification for hanging that infernal weapon on me would have been my size."[115] Of course, size didn't really matter when it came to assigning troopers to the larger weapon types.

Pat Christenson recalled, "When I volunteered for the light machine gun, I never considered why he (Sobel) asked for the larger men. I thought at the time, 'Hell, I'm big enough,' not knowing or thinking that the heavy weapons men had to keep up with the riflemen under all situations!"[116] Unlike Christenson, Don Malarkey didn't request to be placed in a mortar unit, but that's just what happened. "I had been assigned to the mortar squad (of Second Platoon) as a gunner, together with Warren H. "Skip" Muck, who hailed from Tonawanda, New York. We became very good friends and towed the mark under our fiery squad leader William J. "Bill" Guarnere. Bill came to the 506" with excellent combative credentials – 'South Philly'."[117] [118] Malarkey, Guarnere, Muck, Alex Penkala, Clarence "Crash" Tridle, Robert T. Smith and Rod Strohl were not especially tall or well built, but they were wiry, athletic and had remarkable endurance levels. Those attributes aided each greatly when they were assigned to the mortar squads of their respective platoons.

"The Company Commander's Hip-Pocket Artillery"
Easy had a core of outstanding mortarmen during the war, but it all started at Toccoa.
Left: Bill Guarnere led the Second Platoon's Mortar Squad. (Photo courtesy of the Guarnere Family)
Middle: Members of First Platoon's Mortar Squad. (Photo courtesy of the Carson Family)
Right: Rod Strohl was a key member of Third Platoon's Mortar Squad. (Photo courtesy of Rod Strohl)

For Malarkey and "Skip" Muck, their assignment to Second Platoon's Mortar Squad inextricably linked the pair, and they became inseparable. They went everywhere together; on pass, to the PX, even to the latrine. They were both affable and loved having a good time. As Burr Smith had observed, Muck's warm personality had its own gravitational pull, and no one was pulled in more than Malarkey. Winters, a keen observer of his men, understood that the pair's close friendship could become an issue in the future. Both troopers had exhibited leadership traits, and Winters knew both would one day likely be separated by promotion. In addition, Winters saw their friendship as an issue due to the possible deleterious effect should one of them be wounded or killed in combat. Winters had no idea how prophetic his intuition would be. As a result, he separated the pair, sending Muck to First Platoon, while Malarkey remained in Second Platoon.[119]

"Shifty" Powers fired the second highest score in the company at the rifle range and like "Buck" Taylor, was awarded a three-day pass. The only problem was that "Shifty" was flat broke, and had no way to get to the only place the homesick sharpshooter wanted to go, back home to visit his parents. As he watched his friend mope around the barracks, "Popeye" Wynn decided to take matters into his own hands. Grabbing his helmet, he fished a five-dollar bill out of his pocket and threw it in the helmet. "Hey fellas," he called out to the other members of Third Platoon. "Shifty's got himself a three-day pass, but he's got no money to get back to Clinchco. Why don't we all chip in and see what we can do?" The helmet got passed around and most were spurred on to chip in 25 cents, 50 cents, a dollar, or whatever they could. Pretty soon "Shifty" had all the money he would need and then some. As "Popeye" handed the helmet over to "Shifty," he fished out the five-dollar bill and with a wink at his buddy, put it back into his pocket. "Shifty" would be heading home for a well-earned break after all.[120] [121]

* * *

With the rifle range completed, the men continued to acclimate to the physicality of the training, and the cadre began putting them through rudimentary jump training. "Shifty" recalled the methodical nature of the initial parachute training at Toccoa. "They don't just throw a parachute on you, take you up, and expect you to jump out of a plane. They train you in steps. First thing you do, they have a mock-up of a plane that you climb into and practice jumping out. It's about two feet off the ground," Powers said. "They teach you jump, land, and roll when you hit. Everybody thinks jumping out of a plane is hard but it's not."[122]

Camp Toccoa possessed several training aids for the would-be-paratroopers. There was a mock airplane fuselage where they were trained in the procedures of loading into the plane, standing up and hooking up their parachute static lines, moving smoothly in close column up through the plane as they 'jumped,' and lastly, exiting the door. There was also a 34-foot tower, which on top of sat a mock fuselage and door from which the trainees would exit, strapped in a parachute and harness with risers. Once they exited the door, they dropped 20 feet to the end of the risers and then slid down a cable that sloped to the ground. This taught the recruits the all-important technique of leaving the door with the correct body position. They also had saw dust pits for the men to practice their PLFs, or parachute landing falls.

There were several training aids constructed at Camp Toccoa to expose the fledgling paratroopers to the rudiments of jumping. Top left: The "Nutcracker" allowed troopers to get acclimated to manipulating their risers while sitting in a parachute harness. (Author's Collection)
Bottom left: A mock aircraft fuselage was constructed to allow troopers to stand in "stick" order and, utilizing proper body position, jump from the door into a sawdust pit. (Photo courtesy of the Carson Family)
Right: The 34-foot tower combined launching oneself out the jump door, the manipulation of risers, and execution of a proper parachute landing fall (or "PLF") in one exercise. (Photo courtesy of Richard Winters via Erik Jendresen)

Quite a few of the men expressed their fear of jumping out of the 34-foot tower. Second Battalion medics Al Mampre and Ed Pepping had a unique way of testing their mettle when it came to the tower. The two "Band Aid Bandits" decided to climb the tower and jump without a harness.[123] Although they risked injury, they overcame any fears they might have had as they plopped into the saw dust pit below. Courage verified, Mampre and Pepping dusted themselves off and headed back to their barracks, satisfied they could measure up.[124] [125] For the remaining men that had issues with heights, the jump tower was frightening, but most, if not all made it through it with little to no problem.

In addition to the tower and fuselage training, the men were introduced to a contraption nicknamed, "the nutcracker." The nutcracker was a series of parachute harnesses suspended from a large, 20-foot-high wooden frame. Running down the center of the wood frame was an elevated wooden foot bridge. Once the prospective paratrooper was strapped into the harness, he learned how to manipulate or "slip" his risers, which would allow him to have limited control over his canopy, particularly when it came to slowing or speeding up, or sliding to the left or right, during descent. The harness got its nickname from the pain it inflicted on a trooper's groin as he sat in the contraption for hours, learning this vitally important skill. The ability to "slip" could and would mean life or death on, or above, the battlefield.

* * *

Of course, the men did have some down time during their time at Toccoa, and these early days became a chance to get to know each other. Burr Smith, having spent time in a military prep school in California, showed off some outstanding skill in the manual-of-arms with the rifle. He would toss his M1 Garand into the air and catch it, rotating it around with martial ease to the amazement of his bunkmates. His excellence at manual-of-arms earned him the nickname, "Sergeant West Point." Dice and card games were very popular. Both Chuck Grant of Second Platoon and Woodrow Robbins of First Platoon loved to play craps and were considered "great crapshooters." Albert "Albie" Blithe, also of First Platoon, was also particularly fond of dice games. Many a fellow trooper lost some of his wages to the affable Blithe, the flaxen-haired Grant and the string bean tall Robbins. Mike Ranney often organized blackjack tournaments, with himself firmly installed as the resident dealer. In fact, Ranney and Robbins used to bank roll each other because "the action was fast," and often, simultaneously occurring. Many of the men were musically inclined. First Platoon's Richard Bray enjoyed playing the harmonica for the others, with Carson playing guitar and Luz banging on mess tins as a substitute for drums. James "Stinky" Coleman, of First Platoon, was a heck of piano player, not that they ever had access to one while at Toccoa.

These diversions helped to reduce the stress the men were under and allowed them to continue to bond. They shared stories of each other's homes and families. Gathering around the jukebox at the PX, the men hummed and sang along to tunes by Glenn Miller, Benny Goodman, The Dorsey's, and Harry James, and passed around pictures of girlfriends and loved ones. Glenn Miller was a particular favorite of Don Malarkey, and he, Chuck Grant, "Skip" Muck, Tom Burgess, Rod Bain, and Joe Toye, would sit with their back to the wall, taking turns loading songs on the jukebox. According to Don, Bing Crosby's 'White Christmas' made all of them homesick, but they still wore the record out playing it and many others, repeatedly.[126]

The men took every opportunity to blow off steam and relax in order to deal with the unrelenting pressure that Captain Sobel heaped on them.

Left: Dick Winters relaxes on the mattress pad from his officer's bedroll on a small grass field near the Easy Company barracks at Camp Toccoa. (Photo courtesy of the Richard Winters Collection, AHEC)

Middle: Several men formed impromptu musical bands while at Toccoa. Here, Bill Guarnere strums a guitar while James "Punchy" Diel watches from the barracks door. (Photo courtesy of the Carson Family)

Right: At the end of the day, leaning on each other for support was the best way to combat stress. Here, Don Malarkey and James "Moe" Alley pose for a photo at Camp Toccoa. Both wear the HBT fatigues commonly seen during this period of Easy's history. (Author's Collection)

On the weekends that they did earn passes, the men would head into the local towns, hunting for drinks and for many, female companionship. One of their favorite stops was a local establishment called the "The Old Wagon Wheel," which sat just a short walk, or cab ride (using the local 'Safety Cab Company') from the front gate of the camp. Some men even tried to push the limits of a weekend pass by traveling to places like Gainesville or Atlanta, but public transportation was unreliable, as was hitchhiking. Still, many just needed a break from the constant pressure and the exacting standards of not just the Army, but of Sobel as well. There quite a few men who enjoyed drinking to alleviate the stress, but in Pat Christenson's estimation, it just wasn't worth the effort to really tie one on, especially if you traveled far away from Toccoa. "It wasn't worth it – the race to get back to camp, the lack of transportation, the worn out feeling of not enough sleep and too much booze made the following Monday training schedule sheer misery. You'd swore you'd never do it again, but you did (anyway)."[127]

Floyd Talbert, a handsome Third Platoon trooper from Kokomo, Indiana, First Platoon man Robert "Rip" Van Klinken, a Dutch-American from Washington state, and "Gordy" Carson, a gregarious New Yorker also from First Platoon, were more successful than most others in finding local girls to spend time with. Others, like Lipton, Taylor, Perconte, Martin and Campbell T. "CT" Smith were already engaged or married and accepted the risk of trying to get home and back again before their passes expired. Still once they arrived back at Toccoa, they knew that come Monday morning, training would start up again in earnest, no matter what their condition.[128]

Collectively the men, even though most were just boys, had a wicked sense of humor. They routinely played pranks on each other. Floyd Talbert was a frequent victim. "Tab" had a very short fuse when it came to having jokes played on him and his buddies in Third Platoon took advantage of that to its fullest, none more so than Gordon and Paul Rogers. The pair used rhymes and poems for their verbal taunts. "Tab" was deathly afraid of snakes and spiders, and every so often the men would place one of them on "Tab's" cot or in his musette bag. "Tab's" reactions were to the extreme, to include using his rifle butt or entrenching tool to smash the offending creature. But creepy crawly creatures weren't the only thing that provided entertainment when it came to "Tab." Walter Gordon, when rummaging through "Tab's" personal items for joke-playing material, came across a love letter written to him by a girlfriend. As Walter perused the letter he came across two words that would be used to good-naturedly torture "Tab" right up until his passing. The young lady had taken to calling "Tab," "Bunny Duck"...and so would the men, attaching it to catcalls whenever they wanted to get a rise out of him. This ribbing provided entertainment and allowed the men to blow off steam, which was sorely needed.[129] [130]

* * *

Still, Colonel Sink knew how to build morale, especially with his junior officers. Sink did not want the enlisted cadre at Ft. Benning training his officers and senior enlisted troopers, so in October 1942, he arranged to have them jump qualified while they were still at Toccoa. Army leadership at Toccoa contracted a local earthmoving company, R. G. LeTourneau Inc., to clear the top off a small mountain near the base which would later be referred to as "Dick's Hill." The clearing would be used as a runway for the C-47s that would carry the initial sticks of 506th troopers to go through jump training. It was a difficult drop zone to jump into as it was small and bordered by trees and high-tension wires, leaving little room for mistakes.

To determine who would be the lead jumper in the first stick, Colonel Sink held a junior officer 'Olympics' with the winning officer designated the lead jumper. Competition was fierce for the honor of leading the first stick of 506th jumpers. The 'Olympic' event consisted of the following exercises: best time up and down Currahee, most pushups, most chin-ups, and the best time through the obstacle course. Dick Winters won every event except for the six-mile run up and down Currahee. His legs had cramped up during the run, and Wally Moore beat him to the finish line. Despite the loss of that event, Winters ended up winning the 'Olympics' based on points and had the honor of leading the first stick out the door. The men would conduct their five jumps over the course of three days. They would jump twice on September 29th, two more times on September 30th, with the final jump occurring on October 1st. Start time on the 29th would be 0600 hours.[131]

After taking off, the aircraft circled the drop zone five times before disgorging its troopers. Clarence Hester remembered the experience of that first jump quite well. "Some of the officers, while at the Toccoa, qualified as parachutists by packing their own chutes and taking off from the airfield, it was a cut off mountain which the equipment manufacturer had his headquarters. To qualify we had to make five jumps and we jumped into a cornfield. My only bad experience was that I went through my lines and ended up with an inside out chute and blew four panels."[132] Winters, slipping to avoid some high-tension wires, almost broke his ankle landing in the same cornfield but after dusting himself off, he joined those already on the ground.

The first stick of officers from the 506th conducting its initial parachute jump. As a result of the junior officer "Olympics," Dick Winters led this stick out the door. (Photo courtesy of the Richard Winters Collection, AHEC)

The same stick pictured above coming in for a landing on the drop zone at Dick's Hill. Corn stalks can be seen peppering the field.Both Clarence Hester and Dick Winters recalled the difficulties that these stalks caused for the jumpers. (Photo courtesy of the Richard Winters Collection, AHEC)

Soon the recently promoted Captain Sobel and many of the regiment's other officers were also put through their paces jumping into the cornfield. That was until an accident occurred. While the takeoffs were fairly routine, it was the landings that were white-knuckle affairs due to the shortness of the airfield. The jump training was literally put to a screeching halt when the one C-47 the 506th was jumping from, of which then-Major Robert Strayer was a passenger, skidded off the short runway and was disabled. As a result, the Army ordered the jumping to halt, as it was deemed too dangerous to continue with the meager facilities available at Toccoa. Regardless, many the officers had met the requirements to be awarded their wings.

Colonel Sink, not one to pass up a chance to increase morale and *espirit de corps*, held a Prop Blast Party for his newly qualified paratroop officers on November 14, 1942. Sink had commissioned the design and construction of a drinking vessel that would come to be known as the "Sink Grail". The center piece of the Grail was the cup, which was the hub of a C-47 engine. With a base made from wood indigenous to Toccoa, and welded handles made from reserve parachute "D" rings, the 'cup' had one of Colonel Sink's jump wings attached to the front, along with the regimental name etched on the surface. During the Prop Blast Party, the officers, except for the tee-totaling Winters, consumed alcohol procured from the Officer's Mess, partying until the wee hours of the morning. But for the unqualified enlisted troopers, they would have to wait until they arrived at Ft. Benning to get those silver wings and their Prop Blast Party.

With jump qualifications out of the way, the men went back to training just as they had before. On October 7th, members of Easy were crossing over the small creek that flowed through part of the camp's obstacle course. Winters was in front leading the company across the pipes that were suspended above the creek when he lost his balance and fell into the water. He emerged from that water filled with embarrassment and anger over his own failure to negotiate the obstacle properly. He suffered additional indignation later when he found out that he had to buy each officer in the battalion a drink, the standard price for hitting the water.[133]

Nine days later, the company took part in another Friday night march. Sobel laid out a course that stretched for 18 miles. The men were feeling it that night, many nursing minor injuries, Winters included. When one of his men began lagging, he took the man's pack and rifle from him so that he could keep up. By the time they arrived back at camp, Winters' knee had swollen from the pace and additional weight he had carried for much of the march. He ended up staying in camp all weekend to nurse his knee back to health. Two weeks later the Friday night march distance was lengthened to 27 miles. Winters knee was up to the task despite the fact he had to physically support two of his troopers over the course of the march. Winters' personal leadership and overall concern for the men's welfare continued to keep him at the top of the men's collective 'good book.'

As the days slipped into November 1942, the company was put through more focused training on squad, platoon, and company in the attack and defense. The men were provided a single can of stew, several dog biscuits and a single canteen of water to sustain them for a three-day field problem on Thanksgiving weekend that would culminate with the infamous, "hawg innards problem." Thinking that it would toughen up his green troopers to the horrors of combat, Major Strayer arranged to have several truckloads of leftover viscera from a local meat packing company delivered to a makeshift barbed wire assault course on the training grounds. The leftover parts from hogs and cattle were spread across several lanes that the troopers would have to low crawl along. The barbed wire was strung about a foot off the ground and 30 caliber machine guns were positioned to fire over the men's heads. This forced the men to get up close and personal with blood and guts.

According to Walter Gordon, "It was not unusual for us to go through obstacle courses whereby they had spread a certain amount of wire and obstacles in our way and would shoot machine gun fire our heads (while) we crawled through this designated area. The machine guns were fixed on a bracket whereby they could traverse but could not elevate or depress. So, there was a certain amount of safety there provided you didn't stand up and elect to leave the area vertically. If you did that, you got killed (by the machine gun fire). Anyhow, one night in particular, I recall that Colonel Sink or somebody (Strayer) devised this very quaint exercise where they had spread hog entrails all over the ground, and it was a night operation with machine guns (and) detonated charges at specified points which would go off near you in a harmless fashion. And you had to crawl through this vile mess and someone in their infinite knowledge thought this would kind of amount to warfare."[134] Showing his comedic side, Donald Hoobler wrote this about the exercise, "They say we have to have a lot of guts, so they make us crawl through them."

While it is debatable that this exercise hardened the men to the horrors of combat, one effect could not be debated...to a man they were thoroughly disgusted and thoroughly filthy. One E Company officer who believed in the theory that the exercise would harden them was Winters. "For a basic infantry training exercise, this set up a real combat atmosphere. We had a real incentive to keep our head and butt down in the hog guts. I thought it was an excellent exercise, and it's one everyone remembers." Winters had more insight into the reasoning for the training than the rank-and-file. Prior to the field problem, Sink had gathered all the officers together and told them that he wanted the Regiment ready for combat by February of 1943.[135]

* * *

To graduate from boot camp, the Regiment had to take a Department of the Army-ordered physical fitness test. Considering the overall physicality of the training at Camp Toccoa, most, if not all the men in Easy Company felt no pressure in terms of passing the test. So, after completing their 13th and final week of training, the day of the test arrived. The company was broken up into small groups of four men, each overseen and graded by an NCO from a different battalion of the 506th. In the words of First Platoon machine gunner Woodrow Robbins, "we didn't know him from sour owl shit."[136] In theory this would prevent favoritism or bias on the part of the grader.

The test included eight events with each worth ten points. Additional points could be awarded for exceeding the Army-established standard. The men were put through the obstacle course, push-ups, standing broad jump, pull-ups, scaling a ten-foot wall for time, a 50-yard duck walk and finally, a one-mile run. The men ran all the events with little to no time in between each and while the events individually weren't a great challenge, doing them back-to-back was a true test of endurance. At the completion of the test, "Gordy" Carson, George Rajner, Carwood Lipton, Pat Christenson, Lavon Reese and Robert Van Klinken were the top finishers for Easy Company with Christenson scoring the highest point total at 163.[137] Suffice it to say, Easy passed the test with flying colors.

The officers were not exempt from the fitness test, but most passed with ease. Although when it came time for Sobel to take the test, it became an event of great interest to the men of Easy

Company. His young charges gathered around when it came time for the company commander to execute his push-ups. Sobel, for as long as the men had been doing PT with him, had always stopped at the count of 20. The Army test called for a minimum of 30. Wagers were placed, with Johnny Martin making book. Would Sobel pass the test?

Tipper was part of the distant circle of E Company troopers watching the test. "It was public and fair. I was part of a not-so-casual audience perhaps 50 feet away. At 20 push-ups, he was noticeably bushed but kept going. At 24 or 25, his arms were trembling, and I think he was turning red, but slowly continued. How he managed to complete the 30 push-ups I don't know, but he did. We were silent, shook our heads, and this time did not smile. Sobel did not lack determination."[138] [139]

* * *

For quite some time, quite a few troopers with excellent leadership traits had been filling non-commissioned officer positions in the company, all the while unfairly drawing pay as privates despite the fact that they had been performing the duties of corporals and sergeants. Finally promoted in November 1942, many of these men would form the nucleus of Easy's leadership that contributed greatly to its success in combat. Men like the fiery Bill Guarnere, the taciturn "Buck" Taylor, Carwood Lipton, Elmer "Moe" Murray, wiry Second Platoon man, James "Punchy" Diel, former Naval Prep-man Terrence "Salty" Harris, and a Sobel-favorite, William "Stanton" Evans, were all promoted to corporal as they neared the end of their training at Toccoa. While leadership skills were a key, physical ability was also a determining factor. "Buck" Taylor said that the reason he was promoted to corporal before leaving Camp Toccoa to attend TPS (The Parachute School), was because of his superiority at physical endeavors. This was very much an asset for advancement in the early days of the 506th.[140] In addition to the enlisted promotions, Dick Winters was deservingly promoted to the rank of first lieutenant and given the collateral duties of executive officer, ostensibly making him the company's second-in-command.

During this period, Colonel Sink had read an article in *Reader's Digest* that described a forced march that the Japanese Army had completed that had set the world's record for distance and time by an infantry unit.[141] Believing that his "Currahees" were more than up to the task, he set the wheels in motion for two of his battalions to complete a similar march that would shatter the Japanese record. While the First Battalion would ride by train to Ft. Benning, Second Battalion would road march the 118 miles to Atlanta, and for the capper, Third Battalion would march 136 miles straight to Ft. Benning. By now the men of Easy Company were physically hardened. They had survived "Sink's Muscle College" and put up with Sobel's dictatorship. They were used to strenuous physical activity and deprivations. The march to Atlanta would be the final test before attending jump school and one that would not be easily passed.

"Gordy" Carson effectively captured the early days of E Company after his arrival in September 1942. "Our 13 weeks of training, as I look back on it, has been one mess of countless days. We came here raw, very raw, recruits and go from here soldiers. Days of drill, hikes, bayonet drill, rifle practice, maneuvers, physical drills and then there were rainy days, hot days, cold days that filled our 13 weeks."[142] Carwood Lipton also reflected on the difficulties of the training at Toccoa and its influence on the men as they progressed throughout their time in the service. "Our initial training, which developed initiative, self-reliance, and a high *esprit de corps*...proved to be extremely valuable to us in all our combat situations."[143] Ed Tipper summed up the attrition rate across the months of training at Toccoa. "We had a huge turnover. Some men were replaced, and their replacements replaced the same week. My guess is that during this training, we had an overall turnover rate of 150 percent or so."[144] But they had survived and now had one more test before they could attend parachute training.

5

The Last Step Before Jump School

"They're a rootin', tootin', hell-roaring outfit with a marvelous spirit of comradeship between men and officers."
– The Constitution, Atlanta, GA, dated December 4, 1942

The March to Atlanta, 01-03 December 1942

The men knew that they were headed to jump school at Ft. Benning, but they had to make this long march to Atlanta before doing so. And while they were motivated to get to Benning so they could earn their wings, Colonel Sink's main objective had been drummed into their heads before one G.I. service shoe had hit the pavement. "We were supposed to beat the Japs. The Japs had set some kind of record, so we were beating the Japs," recalled Johnny Martin of the messaging from Sink on down.[145]

The cadre members remained with the company for the march but were scheduled to be reassigned once Easy mustered into the Parachute School. The men had started to outpace their stern taskmasters, and it would soon be time for them to be replaced by the emerging noncoms in Easy. Morehead, Stedman, Forde, and Kudla had done their jobs well, but it was time for them to do the same for another unit, but only after completing this last test. The march was broken down into three days of activity. The men would cover 48 miles each of the first two days and the remaining distance, 22 miles, on the last day. Each company would bring its full complement of men, equipment and armament, which included the larger crew-served weapons such as 60mm mortars and .30 caliber machine guns. The larger weapons were broken down and distributed among their crews, but there was no comfortable way to carry the heavy loads, despite the issuance of thick shoulder pads.

This photo, taken early on during the march, shows the Easy Company portion of the Battalion's formation. The weather hasn't turned sour just yet, but it will in short order. Elmer Murray carries the company's guidon. Captain Sobel can be seen to his right to the rear, followed by Harvey Morehead. First Platoon is just behind them, led by James Benton, Robert Rader, George Luz, and Bill Wingett. (Photo courtesy of the Atlanta Historical Center via Sue Hardy VerHoef)

On December 1st, 1942, Easy Company, along with the rest of Second Battalion, formed up for the long march. "Gordy" Carson wrote a detailed account of the march and according to him, it was a truly miserable experience. The column snaked out of Camp Toccoa at 0730 hours and foul weather greeted them. "There was a bad fog and it began to rain after an hour of marching. The rest of the day was spent in endless mud and rain in the Georgia back hills."[146] The excitement and exuberance of the first few miles on that first day was soon replaced by the cold reality of a long and painful forced march. The grousing and cursing typical of young soldiers soon replaced the first few hours' worth of singing and harmonica playing.

Clothed in coveralls made of thin herringbone twill (HBT) material, the cold quickly pierced the fabric and chilled them to the bone. The hard metal edges of the weapons rubbed skin raw. The impact of the march could be seen on the faces of the men, and all of it was captured in pictures of the event. With bags under his eyes and the strain etched onto his face, George Luz hunkered down against the cold with a distinctive white towel wrapped around his neck to ward off the chill, as he marched along near company guidon bearer Elmer Murray. The men marched for five hours, took an hour-long lunch break, then picked up their weapons and packs and continued for another six hours. At over four miles an hour, it was a punishing pace.

The *Atlanta Journal-Constitution* newspaper covered the march extensively. They even went so far as to assign reporters to accompany the marching soldiers for long stretches. Easy Company was featured heavily, which included the portraits of several of the men attached to the articles. Dewitt Lowrey, Benjamin "Bud" Stoney,[147] a tough Native-American First Platoon man from California, Burr Smith and George Rajner all made the paper. Another picture showed a pair of machine gun sections, which featured platoon leaders James K. Davis and a very young-looking Dick Winters, aiming their weapons at the photographer. Another photo featured several members of First and Third Platoons gathered around Third Platoon mortarman Clarence "Crash" Tridle. The group hammed it up for the cameraman, teasing Tridle about the condition of his feet after the first day of marching. While most of the men are smiling, the wear and tear of the march can clearly be seen etched on the faces of Robert Rader, Paul Rogers and Lavon Reese.

Left: As noted above, a group of troopers from First and Third Platoons along several company mess men, gathered around "Crash" Tridle to ham it up for the camera. "Salty" Harris, Lavon Reese, Campbell T. Smith, Richard Bray, William Dukeman, Ed Tipper, and Robert Rader can be seen in the group. (Photo courtesy of the Atlanta Historical Center via Sue Hardy VerHoef) Right: George Rajner smiles for the camera during a short break in the march. (Photo courtesy of the Atlanta Historical Center via Sue Hardy VerHoef)

Newspaperman Lamar Q. Ball had this to say about George Rajner: "Another proud one was Private George A. Rajner, of Toledo, Ohio, a tall, solid ex-steel worker, who lugged a 36-pound machinegun, tossed across one shoulder, for the full 35-mile march Wednesday. Usually, a machinegun is passed from one man to another every 50 minutes on a march. But

Rajner handled this one by himself for a full day's march."[148] The accompanying photo showed Rajner with a broad smile across his face and 30-caliber machine gun on his shoulder. The smile would quickly disappear from Rajner's face as the column accumulated miles of distance.

One of the most endearing images to emerge from the march had to do with Dewitt Lowrey. Lowrey, a fearless, rough-hewn country boy according to Bill Guarnere, had befriended a puppy along the march and the image of the tough paratrooper and the small dog carried in his pack played well with the American public and caught the eye of his fellow paratroopers. "They called him 'Alabama,' after his home state. After six months' training at Toccoa, the 506th marched 115 miles (sic) to Atlanta in three days, eclipsing a record for a forced march held by the Japanese army. Crowds cheered them and newspapers featured photos of the long, lean Lowrey toting a fearsome-looking .30-caliber machine gun on his shoulder and a scruffy stray dog named 'Draftee' in his backpack. The others teased Lowrey: "Alabama's trying to get a movie contract."[149] It wasn't just for show though. "Draftee" had been faithfully following them for miles but had proceeded to wear down the pads on its paws, forcing "Draftee" to yelp and leave little bloody pawprints. Lowrey, feeling bad for the poor dog, scooped it up and using a spare pair of socks, tied the dog to his musette bag. As for the machine gun, the .30 caliber was a bear to carry at 32 pounds, and many of the men that carried the mortars and machine guns had been passing them around to share the burden. Lowrey eschewed the practice and shrugged it off saying, "I'll have to carry it in combat, so why should I pass it off now?" The tough training at Toccoa had paid immediate and obvious dividends.

Left: One of the best known and most endearing photos to emerge from Easy's march to Atlanta depicts Dewitt "Alabama" Lowrey carrying a puppy the men christened as "Draftee," in his musette bag, secured by a spare pair of socks. To Lowrey's left is Elmer Murray, and to his right are "Bud" Stoney and Burr Smith. (Photo courtesy of the Atlanta Historical Center via Sue Hardy VerHoef) Right: A second photo of that series provides a more expansive view of the scene. With the weather worsening, Burr Smith has tied a towel around his neck to ward off the chill. (Photo courtesy of Sean Lester)

Finally, as the sun began dipping from the sky, the column meandered off the main track. Unfortunately, encamping for the evening would not be the refreshing pause the men hoped for. Wind swept across the open bivouac area, in fact it was so strong, it became difficult to keep the camp's fires burning. This included the Battalion Mess Section's field stoves. The cooks were only able to get enough of the stoves lit to provide hot coffee. Instead of hot rations, the men were fed white bread smeared with butter and jam. Some had squirreled away candy bars for just such an occasion. Walter Gordon had stuck a Whitman's Sampler box of candy in his musette bag, and he and his tent mate consumed the entire box that night.[150]

Being early December, even in Georgia, the nights got downright freezing. Staff Sergeant Forde, the cadreman assigned to Second Platoon, not wanting to carry a heavy load, declined to pack his sleeping bag. That decision came back to haunt him that night. Fortunately for him, Winters offered Forde his sleeping bag. Winters fell asleep on top of several barracks bags that had been brought up by the Battalion Supply trains.[151] Rod Bain remembered taking his service shoes off that night and come morning time, couldn't get the frozen leather shoes back onto his feet until he had let the laces completely out. In fact, he vividly recalled the nights during the march, "We had slept on the hard frozen, orange colored, clay soil, putting two shelter halves together to form a small pup tent barely big enough for two troopers."[152]

During a halt for the night, the Battalion mess set up stoves and water heaters. Based on the smoke, the wind has already picked which undoubtedly contributed to a cold night of sleeping on the ground for the men of Easy Company. (Photo courtesy of the Atlanta Historical Center via Sue Hardy VerHoef)

Scenes from the bivouac area. Small campfires provide meager but welcomed warmth from a damp and chilly December Georgia night.
(Photos courtesy of the Atlanta Historical Center via Sue Hardy VerHoef)

Other men recalled hanging their socks up to dry overnight inside their pup tents only to find them frozen by morning. The two meager wool blankets each man carried were not enough to ward off the frigid temperatures and with the men sleeping on the ground, the cold seemed to seep into their bones each night. "Goodie" Guth summarized the weather for all when he said, "(It was) probably the worst weather you could find in that area...ice and snow and wind."[153]

The next day, the high winds continued to blow, but thankfully, the previous day's storm showers had disappeared, and the sun had come out. Still, the unrelenting pace continued. Stepping off at 0930 hours, the column pressed on to Gainesville, only breaking for food at 1500 hours. The respite lasted for just over an hour, and then the men had to fall back into the column to start again. They pressed on for another five and half hours, arriving at their new bivouac site by 2145 hours. By the end of December 2nd, the physical toll of the march began to affect the men.[154] The men were constantly cold. Their socks and service shoes were always soaking wet, either from rain or perspiration. Those, along with their coveralls, were covered in sweat stains, and stiffened at night when the temperatures dropped. The skin on the soles of their feet were rubbed raw and blistering, and their hands had become wrinkled and pruned from being wet all the time.

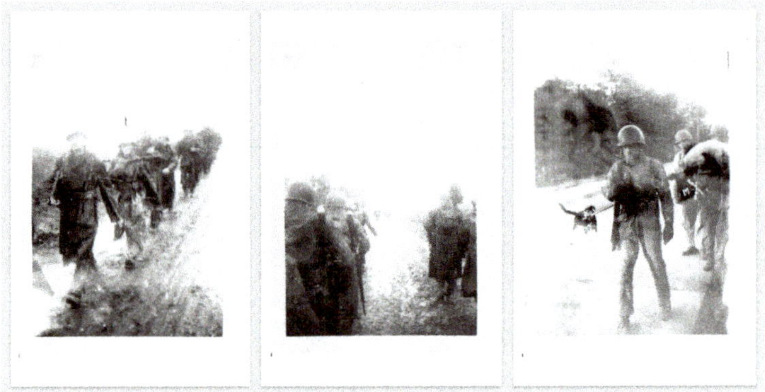

"Gordy" Carson was one of the few Easy men to bring a camera on the march. His photos captured the appalling conditions on the march. In the right photo, George Luz can be seen in the foreground while Elmer Murray attempts to wrestle the guidon pole away from one of the dogs that kept pace with the column. (Photos courtesy of the Carson Family)

The weather worsened each day of the march as did the condition of the men. It became an exercise of mind over matter as the men displayed the mental fortitude to endure the terrible conditions. (Photo courtesy of the Atlanta Historical Center via Sue Hardy VerHoef)

The evening breaks didn't halt the suffering. Men like Joe Toye had to forage for water and wood to make bonfires for the bivouac site even after a full day of marching.[155] Many of the men suffered injuries on the grueling march. "Punchy" Diel of Second Platoon had been plagued by a charley horse in each leg for 15 miles.[156] Don Malarkey was so hobbled by stress damage to his shins, that by the second day he was forced to crawl to the mess line to get his chow. Seeing Malarkey crawling was too much for the ever-reliable "Skip" Muck. So, he grabbed Don's mess tin and told him to head back to his pup tent. Muck then picked up both of their food and coffee rations and came back to the tent to eat with his friend. That small gesture demonstrated the depth of their friendship and the strength of "Skip's" character.[157] [158]

By the morning of the third day, the men were walking zombies. Bill Wingett, who marched near the front of the Easy Company formation, remembered the march well, "The march into Atlanta at the end of basic was also a grueling deal. It wasn't pleasant being in the rain and the snow. At night, we slept in our fart (sleeping) bags and kept our heads covered. You just put one foot in front of the other and got it done, and eventually we made it to Atlanta."[159] Paul Rogers put it succinctly when describing the march, "If it sounds like a lot of miles, it was."[160] Every platoon in Easy outside of the Third Platoon lost men to injury or dropping out. For those injured, they would be allowed to heal, but for those that quit, like those troopers at Toccoa who went to W Company, they would soon be dismissed from the 506th. Still the column trudged along, edging ever closer to Atlanta.

Of all the photos taken during the march, this one is the most well known. Troopers from Easy's First and Third Platoons gather to "huzzah" the photographer. (Photo courtesy of the Atlanta Historical Center via Sue Hardy VerHoef)

They plodded along each day, putting one foot in front of the other. Company Executive Officer Clarence Hester marched at the rear of the Easy section of the column, policing up stragglers. Yet despite the deprivations and monotony, there were diversions. George Rajner continuously did push-ups with his .30 caliber machine gun. Albert Blithe suffered from a toothache, and his complaints kept everyone in stitches. Rod Strohl broke from formation to buy candy for everyone in Third Platoon's Mortar Squad. For his kindness, he was told by First Lieutenant Winters that he would be restricted to camp after arriving in Atlanta.[161] Finally, after covering the final 22 miles to Oglethorpe University, including double-timing the last mile, the men erected their pup tents near the bank of Silver Lake. Except for Strohl and a few others, much of the company was issued passes into Atlanta. But before being cut loose to find decent food, alcohol and women (not necessarily in that order!), the men had to take part in a parade down Peachtree Street.

While many of the troopers grumbled over having to take part in the parade, once it began, they all got into the spirit of the march. Crowds lined the route and cheered and clapped for the would-be paratroopers. Chins were held high and, despite the pain from injuries incurred over the past three days, steps were crisply executed. The thought of a hot meal and some rest, along with music provided by two local high school bands, fueled the men through the Peachtree, Baker and Whitehall streets. The Red Cross met them at the finish line with fresh doughnuts, coffee, peanuts and cigarettes, which was much appreciated by all. At the conclusion of the parade, the men were given three days off before the train ride to Ft. Benning and parachute training.

While exhausting, the men viewed the entire event as Carwood Lipton described it, as "part of the adventure...it was just part of Army life, and I accepted it." Bill Guarnere felt it was all part of Sink's weeding out process. "He (Sink) was going to find out what kind of men he really had, whether they were boys or men."[162] Sink had this to say about the march, "It has always been the maxim of the infantry that the American infantry soldier could always take one more step forward and fire one more shot. The action of these men proves that they can walk one more mile and fire one more burst."[163]

For Rod Bain, the long march had been another formative experience, and the train ride allowed him to reflect on the experience. "We boarded the train in Atlanta, bound for Fort Benning and jump school, our aching feet reminding us of what we had accomplished. The clickety-clank sound of the train soon lulled us to sleep as we made slow progress toward Benning. The train pulled into Benning and thanks to Lieutenant Colonel Strayer whose feet were in the same shape as the lowest private, gave us a reprieve from training so everyone could basically recuperate."[164] For many of those worn out Currahees, the three days reprieve went by way too fast, and before they knew it, they were back it, training, this time, to throw themselves out of perfectly good aircraft.

6

Parachute Training - Earning Those Silver Wings

"You were asking yourself, 'What the hell am I doing here?'"
– Richard Winters

"We was proud of our boots. We was proud of our shoulder patches and we was proud to be paratroopers."
- J.B. Stokes

Lawson Field, Ft. Benning, GA, 04 – 26 December 1942

Located near Lawson Field in an area affectionately referred to as "The Frying Pan", the Parachute School, or TPS, was divided into four stages, lettered A through D, each one progressively leading up to the five qualification jumps. "A" Stage was built strictly around physical training with the goal of getting the fledgling paratroopers accustomed to the strenuous activity associated with manipulating the parachute during the drop. "B" Stage was focused on executing practice jumps from the 34-foot tower. Out of all the hurdles, both mental and physical, that the men had to endure, many cited the 34-foot tower as being the most daunting. "C" Stage was essentially a graduation from the smaller tower to conducting mock jumps from the 250-foot towers. This also included several nighttime tower jumps to acclimate the troopers to landing in the dark. Lastly, "D" Stage was where the troopers packed their own chutes and conducted their five qualification jumps to finally earn their "Silver Wings."

At Fort Benning, the men of Easy Company were finally faced with qualifying for the paratroopers. Everything beforehand was meant to get them to this point.
Left: The 250-foot tower. Several men who had started out at Camp Toccoa and had endured everything thrown at them, requested reassignment rather than take part in this exercise. (Author's Collection via the U.S. Army Signal Corps)
Right: The men would also have to learn how to pack their own chutes. Paratrooper trainees at Fort Benning are instructed on how to insert the chute into the pack tray assembly on the long parachute packing tables. (Author's Collection via the U.S. Army Signal Corps)

When Second Battalion arrived at TPS they were in pretty rough shape physically. Many had not recovered from the ordeal of the march from Toccoa. Don Malarkey was in particularly bad shape, having developed shin splints from the hard marching. First Lieutenant Winters, seeing the obvious pain that the Second Platoon mortar man was in, ordered him to report to the school's infirmary for treatment. He remained there for three days. On the first day, despite being weak from his leg injuries, Don went to the mess hall to get breakfast. When he held out his mess tin, the mess man dropped a hot cake in the tin. Don said, "the damn thing almost bent the mess tin over, it was so heavy." He decided right then that he wouldn't be coming back for breakfast if that was what they were serving. Knowing that Don liked a good cup of coffee, for the two remaining mornings, "Skip" Muck would take his canteen cup to the mess hall and fill it with coffee. Don lived on that coffee until he was released from the infirmary.[165]

The rest of the battalion was granted a three day furlough and ordered to report back on December 7th. Dewitt Lowrey, knowing that he couldn't possibly care for him, gifted his dog "Draftee" to a group of Army nurses who lived on Benning. Other troopers took the time to take stock of their new surroundings and what they saw pleased them. According to First Platoon machine gunner Bill Wingett, "Fort Benning was a luxury place compared to Toccoa. We had big brick barracks and beautiful big mess halls. The Post Exchange had everything you could imagine...the PX at Camp Toccoa had been only cigarettes and chocolate bars. The training at Benning consisted of double-time everywhere you went. We had the parachute mock-up towers. We also had these at Toccoa, except for the parachute device that drew you up and then released you. The highlight of Benning, of course, was the jumps. I still remember the exhilaration of those jumps. That was the frosting on the cake."[166]

The men were issued herringbone twill blouses, pants, caps and brand-new coveralls as well as the coveted Corcoran jump boots.[167] In addition to the new gear issued to them, the men also received regular passes on the weekends. As a result, many of them made trips to Columbus. Some even ventured to Phenix City, Alabama, across the Chattahoochee River. Robert Rader categorized the town as a real, "war zone," with, "shootings, cuttings, clubbings, you name it – it goes on there!"[168] Even with the risks with going out on pass, with Toccoa behind them, the men of Easy Company felt like they could finally see the light at the end of the tunnel. The silver wings were almost in reach, but they would have to go through the four jump school stages before earning them.

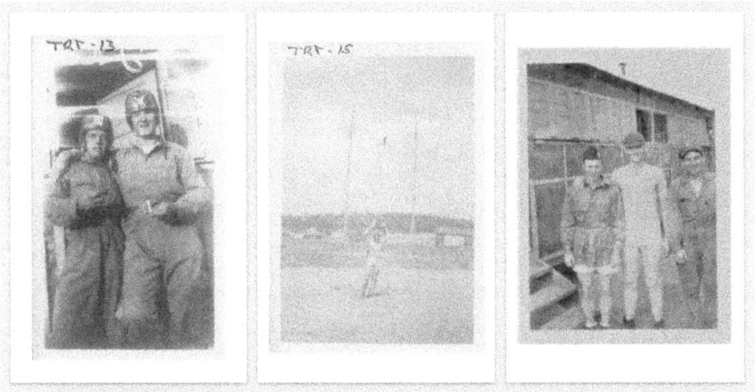

Left: "Bunny Duck" Talbert and "Smokey" Gordon pose with fingers crossed for good luck. Both wear HBT coveralls and Riddell football helmets. In a bit of gallow's humor, Gordon has taped a skull-and-crossbones to his helmet while Talbert has a photo of one of his girlfriends affixed to his. (Photo courtesy of Richard Winters via Erik Jendresen)
Middle: Dick Winters poses in one of the harnesses at the 250-foot tower. (Photo courtesy of Richard Winters via Erik Jendresen)
Right: It must be laundry day. "Goodie" Guth, "Smokey" Gordon and "Crash" Tridle pose in various states of dress outside the barracks. (Photo courtesy of Forrest Guth)

When the men arrived at "A" Stage, they quickly found out it was going to be a test of wills...them versus the school's cadre of instructors. Made up noncoms that had previously completed jump school, the instructors had read the newspaper articles about the march from Toccoa and were not impressed by the Sink Men. As far as they were concerned, the 506th hadn't really accomplished anything. So, the cadre developed a nickname for the Regiment to let them know where they stood in the pecking order; "the Walkie-Talkie Non-Jumpie outfit." Their next order of business was to break the students down physically and that would occur in "A" Stage. Unfortunately for the cadre, things did not go as planned.

Since the company's officers had already qualified as paratroopers, they spent their time at Ft. Benning taking classes which included learning to swim, riding horses and motorcycles, and more military-focused classes, like demolitions. The demolition course was especially memorable. Winters along with Wally Moore, and Frank Reis, the Battalion's Light Machine Gun Platoon Leader, spent several days traveling to local water and power plants, telephone exchanges, and radio stations to learn how to use explosives to disable or destroy similar facilities. They also learned how to blow up power lines, sabotage trains and rail lines, bridges, and cars. Instructors also taught the eager students how to use explosives to penetrate concrete and steel and knock down trees to use as impromptu barricades. Classroom time was spent mastering the mathematical formulas to correctly estimate how much explosives to use for each task.[169]

While the officers went through their additional training, the enlisted troopers were turned over to the TPS instructors who were all seasoned noncoms. The instructors decided early on that they would attempt to intimidate and break the 506th troopers with physical activity. What they hadn't planned on was that those troopers had just graduated from "Sink's Muscle College." For the Toccoa Men, the challenge wasn't all that difficult due to the tremendous physical shape they were in. This was especially true for the men of Easy Company, where Sobel's strict training methods paid off handsomely.

In fact, it was the so-called victims that ended up tormenting the instructors, who just couldn't physically keep up with the troopers. In one instance, Pat Christenson was ordered to "drop and give me 20" push-ups by an instructor. When Christenson got down, he looked up at the instructor with a grin and asked if he wanted them "one-armed or two, because I can do this all day." The instructor, incensed with the insubordination and not wanting to be shown up in front of an entire company of student paratroopers, told him, "One-armed...and make it 40." Pat knocked them out with ease, and adding to the insult, immediately switched to the other arm and began knocking out an addition 40 push-ups. The instructor could only stare at him, red-faced and embarrassed.[170]

Two Toccoa stalwarts and a Benning new join.
Left: Elmer Murray was an early noncom in Easy. Highly intelligent, he became the company's Operations Sergeant.
Middle: Phil Perugini, tough-as-nails Third Platoon trooper, dressed "to the nines," in his M42 jump shirt, dress shirt and tie. (Author's Collection)
Right: William "Willie" Wagner joined Easy while at Benning. He would become a Third Platoon mainstay during his tenure with Easy. (Author's Collection)

Third Platoon man Phil Perugini did much the same to another TPS instructor. According to Doc Pepping, the instructor ordered Phil to do 40 pushups as punishment to make an example of him. Instead of the 40, Phil knocked out one hundred push-ups. Phil continued to torment the cadre after that incident. Another TPS instructor targeted Phil and ordered him to knock out some push-ups as a punishment. Instead of doing two-armed push-ups, Phil cranked out the required number of push-ups with one arm. As before, the instructor could only stand there embarrassed. The tormented, again, had become the tormentor.[171] [172]

A second and final embarrassing incident would spell the end of "A" Stage for the entire Regiment. While the reason for the incident remains cloudy, the result could not have been clearer. At some point, early on in "A" Stage, one of the instructors had his fill with the men of 506th. With several of his fellow cadre members present to assist with the effort, he decided to take a company of 1st Battalion students out and run them into the ground. The attempt backfired badly. The run lasted the better part of a couple of hours and the instructors, unable to keep up with the Sink Men, quickly winded themselves. As they lay along the way, the troopers continued circling the track like birds of prey, mocking the instructors all the while. When word of this episode got back to the TPS leadership, they called Colonel Sink to the school's offices. After a brief discussion, it was agreed upon that the 506th required no additional physical conditioning and would bypass the remainder of "A" Stage. The unapologetic, cocky troopers were quickly headed to "B" Stage.

After their arrival at Benning, one thing that quickly became apparent to the men of Easy Company was that being a paratrooper was as much a mental challenge as a physical one. There was a certain frame of mind that had to be maintained and one thing that helped to hammer that point home was the use of double-timing. As Bill Wingett stated they, "Needed to be moving at a paratrooper-type pace, no matter what they were doing or where they were going."[173] As a new officer, Dick Winters had been impressed by several paratroopers that he encountered before he joined the 506th. He saw motivated, prideful troopers..."good looking men"...one's with great spirit, who didn't slack off or take the easy way out. When they moved out, they did so with purpose. And double-timing everywhere they went was a physical manifestation of that mindset. Winters wanted to be part of that type of unit.[174] They all did, and it was time for the rubber to meet the road.

TPS was not the first time that Easy Company had been exposed to parachute harnesses, towers and mocked up aircraft fuselages. The parade ground at Toccoa possessed a 34-foot tower and a mock-up of the C-47 aircraft for the men to train on. The men spent a considerable amount of time jumping out of Toccoa's aircraft mockup and tower, landing in saw dust pits. "B" Stage became an extension of that initial training. The also practiced on the nutcracker again, a contraption that they were well familiar with from Toccoa.

The men also conducted training jumps from the 34-foot tower, which was something that most, if not all of them experienced at Camp Toccoa. Each trooper ascended the tower and was strapped into a harness. The harness risers were then attached to a cable that led to the ground. After a swat on the butt, they leapt out, dropped 15 feet until the line caught, then slid the remaining distance along the cable to the ground. Many of the men were excited to have made it that far in training. Some, like Joe Toye, unknowingly threw caution to the wind when doing practice jumps. As a result, their body position upon exiting the tower was less than perfect. Toye explained, "I guess I was so excited, I just jumped. Never figured body position, or something, and got told about it when I got down to the ground. Somebody said, "Hey, that's not the way we do it here!" Toye obligingly knocked out the punishment of 50 push-ups and then ran to get back in line.[175] Even with a few instances like that, by in large, the training went off without a hitch.

Rootin' tootin' paratroopers. "Goodie" Guth, Ken Baldwin, John Connell, and Phil Perugini pose in jump suits, field gear and M1 rifles with bayonets while at Fort Benning. (Author's Collection)

According to Robert T. Smith, the men ran to and from training, all told 4 miles round trip. He recounted the training in detail. "Last week our training consisted of chute packing in the morning and mock door, mock tower, landing trainer, tumbling and exercises. This week we will continue chute packing and work on the high towers."[176] After a week of cutting their teeth on the nutcracker and 34-foot tower, the men progressed to "C" Stage. That was where things began to get serious due to the presence of the daunting 250-foot tower. While most of the men had no issue with traversing the 34-foot tower, the 250-foot tower was a whole other animal. Those troopers with a fear of heights found the tower absolutely hair-raising, and if they didn't pass "C" Stage with its seemingly mile-high tower, there was no way they would graduate.

Rod Bain recalled the 250-foot tower, not for the height or for the drop, but for how he injured himself upon landing. "I remember spraining an ankle on Benning's 250-foot freefall (sic) jump tower, coming backwards as I approached the ground. While the rest of E Company continued heavy physical exercises, I had only to report to sick call each day for a week and soak my swollen ankle in a whirlpool bath."[177] Bain eventually returned to full duty, but for those who refused to make the drop, the results were swifter than the descent from the tower.

Several troopers willingly quit and were subsequently transferred to the School's Casual Company to be out-processed. Some were even so scared that they went AWOL or absent without leave and were later rounded up and transferred to the regular infantry. Between shipping out and new arrivals, from 1 January to 28 February 1943, all told, Easy Company lost twenty troopers and gained nineteen. So, while the company's end strength remained somewhat stable, the tumult over the transition of troopers stayed high.

While those men were leaving, others were arriving to replace them in the ranks. Soon-to-be company stalwarts William "Willie" Wagner and William "Bill" Kiehn joined Easy at TPS. For "Buck" Taylor, his first few encounters with Bill Kiehn were particularly memorable. "(We) got a new recruit (Bill Kiehn) in 3rd Platoon, a kid that didn't want to take orders and thought he knew everything. 1LT Winters came to me and told me to take the kid out for an hour of close order drill starting at 0500 hours every day until he shaped up." And shape up he did. According to "Buck," Bill became "one of the best soldiers in the platoon."[178] Yet despite these transitions, training continued unabated.

One of the training implements the men had to endure was the "Windcharger," which was meant to simulate being dragged across the drop zone by strong winds. The trainees were required to maintain control of their risers and collapse their chute. (Author's Collection via the U.S. Army Signal Corps)

Another training device made a memorable impression on the young troopers...the wind machine. Looking like the engine and fan assembly from an air boat, the wind machine taught troopers how to control their parachutes after landing, in case there was significant wind on the drop zone. This also helped the troopers learn how to collapse their chute while being dragged along the ground. This was tough training, tough on the men and tough on their uniforms. The constant dragging on the rough ground, with its ruts, pine needles, rocks and pine cones, tore up their herringbone twill coveralls at the knees and elbows, along with their prized Corcoran jump boots. The men were kept busy repairing rips and holes and polishing their boots in their down time. And even though the instructors were in charge, it still didn't stop Captain Sobel from inspecting the men and their barracks on the Saturday of each week.

High performance on the jump training exercises could and often did result in the issuance of three-day weekend passes by the cadre. According to "Buck" Taylor, Sobel oftentimes tried to block these passes, but the men were becoming more confident soldiers, ones who pushed back on the chickenshit from time to time. Taylor explained, "Several of us (NCO's) had won three-day passes, and I was one of the ones who won a three-day pass, and I had my bride-to-be come down on a train to Jacksonville. I was going to meet her there, go over from Benning. And on our Saturday inspection, he purposely gigged several of the NCO's who had passes, me included, and I think Lipton was another one. And of course, I was very annoyed at that. But I already had the pass in hand, it was issued, so I went ahead and I took the three-day pass, and he never did anything about it when I came back."[179] And with the passing of "C" Stage, the men had no other obstacle in their way of getting their jump wings, except for their five qualifying jumps in "D" Stage.

Other than making their jumps, the primary focus of "D" Stage was the men learning to pack their own parachutes. Coincidentally, the men paid extra close attention to the details during this phase. Since they were packing the chutes they would jump with, each man made sure to do as precise a job as they could to insure a fully blossoming canopy followed their exit from the aircraft. Each morning, the company marched out to the packing sheds near Lawson Field and were greeted by the long packing tables. Stretched out along them were their parachutes. Held down by weighted shot bags, the chutes were held taut, while in small groups, the would-be paratroopers folded the panels in upon themselves. Once that was done, the chute and its corresponding static lines, were packed into the parachute harness pack trays and secured by covers that attached to the static line. The extra static line that hung loose was then secured to the upper sides of the pack tray assembly by rubber retaining bands. The assembled chutes and their reserves were then placed in kit bags and staged, in preparation for the company's first jump, which would take place on 21 December 1942. Four more jumps would take place, with the final occurring on Christmas Eve.

Left: A stick of paratrooper trainees make their way into a waiting C-47. (Author's Collection)
Right: A stick of paratroopers descend to the drop zone. (Author's Collection via the U.S. Army Signal Corps)

On the morning of the 21st, the men again marched out to Lawson Field and the packing sheds, only this time, they drew their kit bags and were divided into 18-man sticks. "Gordy" Carson, Max Clark, "Duke" Dukeman and Taskel Ellis helped each other buddy-rig their chutes.[180] Like the nutcracker, the harness was strapped on to each trooper tightly to prevent additional shock and injuries when exiting the plane. The men boarded their assigned aircraft and sat nervously in the bucket seats. For many of them, this would be the first time flying. After firing up the engines, the pilots throttled up and the C-47 proceeded up the taxiway. Finally positioned at the departure lines, the tower cleared them for takeoff. It was time to earn their jump pay.

Burr Smith vividly recalled that first jump. "(I) made my first parachute jump, an experience I'll never forget! Awfully scared, but so was everybody else so I didn't feel so bad. The sensation of falling through space is indescribable. Just like a dream. The opening shock was slight, but I hit the ground like a ton of bricks!"[181] For most of the men, the freedom of dropping to the ground was exhilarating. "Skip" Muck and Alex Penkala exchanged whistles and good-natured insults as they drifted towards the drop zone. In fact, the sky was filled with troopers calling out to buddies. Most, like Burr Smith, forgot their training in the excitement, and hit the ground hard, which earned them punishment in the form of push-ups from the instructors waiting on the drop zone. They happily knocked them out and recovered. Robert Bloser was the only Easy trooper injured on that first jump, when he twisted his ankle upon landing. Others, like Frank Perconte, admittedly, were "scared as hell", but escaped unscathed.[182]

Left: Descending troopers approach the drop zone. At this point, the troopers are manipulating their risers and preparing to execute a PLF. (Author's Collection via the U.S. Army Signal Corps) Right: Firmly on the drop zone, a paratrooper trainee struggles to maintain control of and collapse his canopy. (Author's Collection via the U.S. Army Signal Corps)

"Shifty" Powers, also nervous about the jump, tried to figure out ways to overcome his fear. "They taught us to yell 'Geronimo' when we jumped. It was supposed to take our mind off jumping, but it didn't work for me, so the instructors told me to yell 'Currahee', but that didn't help either. They finally told me to just yell anything I wanted. Well, I finally found something I could yell and it kind of distracted me, I'd yell 'OH SHIT, WHAT AM I DOING HERE?!?!' It worked for me."[183] With the first of the jumps out of the way, many of the men felt like they could take on the whole German army.

Rod Strohl had a memorable experience on that first jump, but for the wrong reason. "We jumped in coveralls, and I had a pencil in my pocket. And the jump instructor...came up and says (sic)...he grabbed the pencil and threw it at me...he says (sic), 'What's the matter, soldier, you want to kill yourself?' I wish I could have seen that guy when I jumped into Normandy and let him see what you can jump with. (He) thought I was going to kill myself with a pencil in my pocket."[184]

It was the second jump, on the 22nd of December, that really tested the metal of the fledgling paratroopers. With the first jump, they didn't know what to expect. There was no such luxury with the second jump. Again, it was Burr Smith that provided a voice to what many of the men felt. "Living on my nerves - how long can I do it? Parachute jumping is terribly exciting, but I can honestly say that I don't enjoy it. It's (sic) fun after the canopy opens, but that fun doesn't overshadow the fear that seizes me as I go out the door. Lots of the boys thrive on it, but I'm too damned excitable. I may grow to enjoy it later on, but only time will tell."[185] One man in Third Platoon, Joe Ramirez, froze in the door on the second jump. Rod Strohl and Forrest Guth watched Ramirez turn around, walk to the front of the plane and sit back down. In doing so, he put himself in jeopardy of being thrown out of the school and the paratroops. The C-47 circled the drop zone as the jumpmaster went up and spoke to Joe. It was made clear to him what a jump refusal meant, so Joe picked himself up, attached his static line to the cable and when the plane came around, threw himself out the door. As they watched from the ground, Strohl and Guth, clearly impressed by Joe's fortitude, remarked that it took him more guts to do it the second time around.[186]

With Christmas fast approaching, it was decided that the men would stack two jumps on the 23rd in order to allow them to qualify on Christmas Eve. A massive Christmas dinner had been planned for the holiday by Colonel Sink to celebrate the Regiment's accomplishment, but they would have to knock out the last three jumps first. "Gordy" Carson, this time positioned as the lead jumper for his stick, sat near Johnny Martin, watching the ground slide past them through the open door. As soon as they landed, the men were transported back to the airfield where they were issued another chute and reserve. They were quickly taken up and made their fourth jump.

"Shifty" Powers recalled a humorous incident on that jump. Forrest Guth was several troopers in front of "Shifty." When the green light lit up, the men started pushing towards the door. As Powers got closer to the door, he could see "Goodie" lying on the floor, having slipped and fallen. "Shifty" quickly made the decision to hop over him and head for the door. What surprised "Shifty" was that as he landed, he noticed "Goodie" collecting up his chute. "Shifty" had no clue how he had beaten him down to the drop zone. Still, they were down and only had one more jump left to qualify as paratroopers. By the end of the day, the men were back in their barracks, a little sore from the pair of opening shocks and landings but still riding a wave of emotion and exhilaration. They were almost done.[187] [188]

Members of Easy Company mug for a photo before heading out to the airfield for one of their qualifying jumps. Each wears HBT coveralls and Riddell football helmets. (Photo courtesy of Forrest Guth)

Just like the three days prior, the men got up early, marched out to Lawson Field and were issued their chutes. While the weather had been good those previous days, it was even better on the 24th. There was little to no wind on the drop zone. The temperature was comfortable, and the pilots dropped the men right on the mark. This isn't to say that all went as planned. The sticks had been mixed up and "Skip" Muck had to push a faltering trooper out the door. According to Burr Smith, "A man from Headquarters Company froze in the door (just two men ahead of me) but my buddy Skipper Muck kicked him out the door – cruel treatment, but a scared jumper can cause the death of a whole stick if he freezes in the door when we got into action."[189] Once again, the men drifted down to the ground calling out to each other and razzing those who had poor exits. With the exception of Verlin "Sadie" Hawkins, every Easy trooper completed their fifth jump. Hawkins, who froze in the door and refused to jump, landed with his plane and was immediately removed from the company. It was sad for the others in the company to witness as Hawkins had been a member of Easy from early on at Toccoa, but they had all been warned about the consequences of refusing to jump. For the rest of Easy Company, they were now qualified paratroopers.

On the morning of the 30th, the entire Regiment fell out in Class A uniforms for the jump wing pinning ceremony. The men were now officially permitted to blouse their boots. Some troopers used rubber retaining bands from the packing sheds. Others used condoms to get the proper 'mushroom' affect. After the men formed up by company and battalion, the Regimental Adjutant gave the order to for the troopers to be awarded their wings. Lieutenants Matheson, Winters and Moore moved up and down the ranks of their platoons, pinning silver parachutist wings on each of his troopers. Sobel said little to his assembled company beyond a brief congratulation.

HEADQUARTERS
506th PARACHUTE INFANTRY
OFFICE OF THE REGIMENTAL COMMANDER

Fort Benning, Georgia,
December JAN 9 1942 1943

MEMORANDUM TO SOLDIERS OF THE 506th PARACHUTE INFANTRY:

You have now become qualified parachutists and wear the wings of the Parachute soldier.

You are a member of one of the finest regiments in the United States Army and, consequently, in the world.

You are about to go on furlough, into the homes of relatives or of friends.

I feel that I should remind you of certain things that are expected of you—not only while on furlough, but also a [illegible] by which you are expected to govern your life and your actions:

1. You must keep in mind that first you are a soldier in the Army of the United States; that you are a parachutist, the elite of this army, and finally that you are a member of the 506th Parachute Infantry.

2. You must walk with pride and with military bearing.

3. You must be careful of your personal appearance, keeping your uniform neat at all times.

4. You must do nothing to bring discredit upon the Army, Parachute Troops or this Regiment.

5. Remember our battlecry and motto, "CURAHEE", and its meaning: "Standing Alone". We Stand Alone Together.

The Regimental Commander desires that you convey to the members of your family his personal greetings.

R. F. SINK,
Colonel, 506th Parachute Infantry,
Commanding.

"Behave yourselves and don't get arrested!" With their well-earned jump wings pinned, Colonel Sink addresses the 506th before releasing them on furlough. (Photo courtesy of Ray Aebischer)

Once every trooper had been pinned, the Regiment paraded before the command staff, which stood on a raised platform. At the conclusion, Colonel Sink called his men around him and read the Order of the Day, of which each man was given a printed copy. The men were given 10-day furloughs with two stipulations; they were to return on time and most importantly, they were ordered to, "behave yourselves and don't get arrested!" Sink would not put up with any conduct that brought shame and embarrassment to the Regiment. With a smile and wink, the Colonel released his men. The newly minted paratroopers were flying high with confidence, and best of all, most were headed home to see families that many had not seen in over a year.

Scenes from a well-earned furlough.
Left: "Bunny Duck" Talbert, Art "Mad Dog" Youman, Paul "Hayseed" Rogers and "Goodie" Guth get ready to head out on furlough after qualifying as paratroopers. (Author's Collection)
Middle: Standing in the snow-covered yard of his family, "Gordy" Carson poses for a photo wearing his M42 jump suit and jump boots. (Photo courtesy of the Carson Family)
Right: Skip Muck posing for a photo with his girlfriend, Faye Tanner, back in Tonawanda, New York. (Author's Collection via the Muck Family)

7

Easy Company in Transition

Ft. Benning, GA, 05 January – 09 February 1943

At the conclusion of their furlough, the men slowly gathered back at Ft. Benning, starting on the 11th of January. While most of the company was able to slowly reacclimatize to the Army way of doing things, Burr Smith, Rod Bain, Joe Toye and "Rip" Van Klinken immediately received orders to attend a two-week Demolition Specialists Course. The remainder of Easy stayed on Benning, albeit on the "Alabama" side of the base, waiting for the ranks to fill. Many pulled K.P. duty, peeling potatoes and assisting the cooks in preparing meals for the company. To keep the men sharp, Matheson, Winters and Moore conducted hip-pocket classes on tactics, weapons maintenance, and of course, physical training, which included 9-mile runs every other day, except on Sundays. Dust from the red clay around Benning soon covered everything. Still, the men took it all in stride, as conditions were far superior to that of Camp Toccoa.

With a bit more spare time on their hands, many Easy troopers took photos of themselves for their families.
Left: Bill Guarnere, obviously pushing uniform regulations with his M42 blouse unzipped, poses outside a barracks building with his M1 rifle. (Author's Collection)
Middle: Third Platoon radioman, Rod Bain, shows off his hard-earned M42 jump suit and Corcoran jump boots. (Photo courtesy of the Bain Family)
Right: S-4 Supply Clerk, Max Clark, poses in his M42 jump suit, helmet and field gear outside the Easy barracks. (Photo courtesy of the Clark Family)

Winters had really come into his own beginning at Toccoa, through the march to Atlanta, and throughout jump training. He was a fantastic leader of men and many of the E Company men marveled at their luck of having him in the ranks. "Rook" Rader spoke about Winters in almost reverential terms. "This man was openly friendly, appeared genuinely interested in your 'interests' and was heavy on physical training. Dick was a patient man. Also and intelligent man – he was a foreseer-in-the-future man. Dick was easy to know...for an officer he was 'human.' Dick figured if he was to survive during the war, he had better train 'us' guys real (sic) good! Lt. Winters was almost shy – wouldn't say 'shit' if he stepped in it! Dick always had time to help train you on a weekend, and expected nothing but the best from you – at all times!"[190] The more time the men spent with both Winters and Sobel, the stark differences between their leadership styles was laid bare for all to see.

Sobel, on the other hand, continued his martinet ways. Lipton recalled one particular incident in which he was able to change Sobel's mind after he was punished for a so-called infraction. Sobel had decided to grant a three-day pass to any trooper who had not missed a day of training, whether sick or otherwise, during his time in Easy Company. According to Lipton, this amounted to seven or eight men. Before the pass was issued, the company, minus Sobel, went out on a field problem, during which time Sobel inspected the barracks. Every man who had been issued a pass was found to have infractions. After the company returned from the field problem, Sobel announced to the formation that all three-day passes had been revoked due to the infractions.

Lipton had been gigged for not having laces in his jump boots. He quietly seethed. After being informed that he had been awarded a pass, he had called up his wife Joann and arranged for her to come down to Atlanta, where he would meet her for a rare weekend together. With the pass canceled and no way to tell Joann that meeting her was going to be impossible, he went to his platoon leader, Wally Moore, and informed him of his intent to go AWOL to meet his wife. Moore counseled him against it. "Don't do it. It'll always be on your record. Sometimes Sobel will listen to reason. Go in and talk with him." Lipton thought about it. The reason he had no laces in his jump boots was because he had taken them out to shine the boots, and to wax the laces. What Moore said made complete sense. What Lipton had done wasn't some sort of dereliction of duty or flaunting of the rules. It was done to exceed the standard. He requested and was granted an audience with Sobel. After explaining why he had no laces in his jump boots, Sobel ordered him to present the laces as evidence. They were waxed to perfection. Satisfied that Lipton was being truthful, Sobel stared at him, "Okay. Your restriction is lifted." It was a rare case of leniency on the part of Sobel.[191]

By mid-January, the company had largely reformed. Training schedules posted on the bulletin board indicated that weapons training would begin in earnest. Winters was detailed to present classes on the Thompson and carbine on the 17th. In addition, Easy would observe a bazooka demonstration, as well as ones on the British Bren and Sten guns. On January 18th, Easy Company drew their individual weapons and went to the range. Each trooper was permitted to fire every weapon in the company's arsenal. "Gordy" Carson recalled firing, "the M1, sub-Thompson and the carbine", an exercise he considered 'fun', especially firing the Thompson.[192] Robert T. Smith agreed that the firing range was enjoyable. "We were out on the firing range all day firing the rifle, machine gun, tommy gun and carbine; my ears are still ringing and last night every time I lit a cigarette it tasted like I was smoking gun powder. For some reason I surely do like to hear 'a lot of noise.'"[193]

After cleaning their weapons and having them inspected

by their noncoms, the men were given something unheard of during their time at Toccoa...the afternoon off. Most lounged around in their barracks, reading and writing letters. After dinner, many filed over to the base theater to watch 'Commandos Strike at Dawn', starring Academy Award winner Paul Muni. The next day, the 19th of January, Easy went to the field to take part in what was supposed to be an exhaustive two-day problem but returned that same day to attend a Regimental formation. Most had no idea what the formation was for, but they would soon find out.

For many years, it was widely believed that only one man returned late from furlough and as a result, was transferred out of Easy Company because of his violation of Colonel Sink's mandate. In fact, there were 18 men who returned late, including well-known troopers like Rod Strohl, Bill Wingett, Ed Tipper, "Skip" Muck and Third Platoon man Bill Maynard. Many of these late returnees had to deal with an unreliable national travel system. Buses, trains and in a few instances that they could afford it, air travel could not be counted on for arriving at destinations on time. Captain Sobel realized quickly that he couldn't just drum out 18 troopers. He knew that it would create a terrible shortage of troops and would also reflect poorly on his leadership. He also knew that many of those 18 were excellent troopers. So, in the end, he chose to make an example out of one of them; Gregory Rotella, who was transferred out of the parachute infantry.[194] For the rest of the late arrivals, they suffered a loss of jump status for 30 days and with it, the extra $50 in jump pay for that month.

What a difference a few weeks make. The men of the 506th had been flying high since graduating from jump school. Before sending his Regiment out on furlough, Colonel Sink reminded them to stay out of trouble, which included not returning late. For various reasons, many did return late and because of it, several men were made examples of, including Easy's Gregory Rotella. Left: A 506th officer is presented his jump wings before the Regiment was given a furlough. (Photo courtesy of Ray Aebischer) Right: Salve Matheson presides over the "drumming out" ceremony. This affair left a bad taste in the mouths of the Easy Company men. (Author's Collection)

The "drumming out" ceremony as it came to be known was a particularly sad, somber affair, one that Carwood Lipton called, "a most barbaric ceremony that I've ever seen. (One that) really made an impression on our lives."[195] With every company formed on the parade field in Class A uniforms, each man being discharged from the paratroops was put on full display in front of the entire Regiment. "Buck" Taylor, with M3 Grease Gun slung on his shoulder, marched Rotella up to the command staff. They cut the Airborne Command patch off his sleeve, then cut off his overseas cap parachute patch and then finally, ordered him to unblouse his jump boots. He was then told to remove his jump boots and put on a pair of service shoes. This final indignation infuriated many of the men who considered the humiliation to be a "raw deal". Rotella was then handed his barracks bag and with the Regimental Band's drummer tapping out a muffled cadence, he was marched off to be reassigned from the 506th. Many of the troopers who witnessed this thought that after surviving the crucible of Camp Toccoa and then jump school, some sort of leeway should have been given. Instead, these men that had put up with so much were forced to suffer a great embarrassment in front of their friends and peers. But the lesson was learned; Colonel Sink would cut no slack. If you violated the rules, you would pay dearly for it.[196] [197]The following day was a harbinger of what was to come for the remainder of Easy's time at Benning; 9-mile runs, weapons maintenance, short field problems and organized athletics. Morale flagged during this period as rumors of a move to Alabama instead of heading to North Carolina, permeated each barracks building. Promotions continued to be awarded to deserving troopers. Bill Guarnere recalled the almost comical way that he had been promoted to sergeant. "We were out on a field problem, and Evans said to me, 'Say. I knew there was something I wanted to tell you. You're a sergeant!'"[198] Johnny Martin and "Bull" Randleman made corporal around the same time. Randleman, an imposing "big boy," in particular, made an excellent NCO since he was considered "very GI", or in other words, strict and by-the-book.[199] Guarnere excelled in calling cadence, a vital component in leading troop formations, while Martin was quick-witted and skilled in figuring out field problems. These promotions, as well as many previous ones, demonstrated that every key leadership position was occupied by a man who survived the crucible of Toccoa. Easy had grown and would continue to grow its own leaders.

February of 1943 saw a substantial turnover in personnel, including many Easy troopers who had been with the company all through Toccoa. Several had gotten drunk on payday weekend and went AWOL. Punishment from Sobel was swift. Chester Eschenback, Fred Bealke, Jay Dickerson, and many others were sent packing, and newcomers Herman Collins, Cleveland Petty, George Potter, George Yochum, Clarence Howell and William "Pop" Woodcock all joined the company. Of the six, Collins would die as part of the ill-fated Stick #66, and Howell would be injured on the drop on D-Day. Potter made the Normandy and Holland jumps and would later be evacuated from Bastogne with frostbite, finally returning to the company in mid-January 1945 to finish out the war. Petty would be wounded twice, once at Brecourt Manor on D-Day morning, and again at Carentan, during the fighting at Bloody Gulch. Yochum would be transferred out of Easy before D-Day, to Fox Company, where he would serve until the end of the war. But all of this lay in the future.

Despite the upheaval in the ranks, the men continued to train, with Easy conducting a parachute jump on February 8th, its first since completing jump school. "Gordy" Carson, in the #2 position, vividly remembered following jumpmaster Lewis "Blackbeard" Nixon out the door. They also learned how to build rope bridges for use over water or land obstacles. Close order drill was a daily event, and as before, the nine-mile runs were undertaken every other day. It wasn't all work and no play. The men took part in basketball, football and baseball games as well. As much fun as they had with the organized sports, their time in Georgia had finally come to an end.[200]

Left: George Luz and "Bull" Randleman pose with their weapons outside the barracks at Fort Benning. (Photo courtesy of the Carson Family)
Middle: Skip Muck, Don Malarkey, and Jay Dickerson mimic standing in the door in the Second Platoon barracks. All three served in Second Platoon's Mortar Squad at one time or another. (Photo courtesy of Don Malarkey)
Right: Taken at the same time as the middle photo, Skip points out a C-47 in flight to Don. (Photo courtesy of Don Malarkey)

After seven weeks of training around Fort Benning, the men were told to pack their belongings. The planned next step in their training would begin with them being trucked up to North Carolina for eleven weeks of intensive field training. Instead,

Sobel, ever the taskmaster, had them march the eleven miles to their new camp. If Camp Toccoa had been their introduction to the school of infantry, and Benning their apprenticeship, then Camp Mackall would be their finishing school.

8

Becoming Masters of their Craft

Camp Mackall, NC, 09 February – 22 May 1943

After a long, dusty march to Camp Mackall, the men began taking in their new surroundings. The camp was nestled among copses of pine trees that are indigenous to North Carolina. It had originally been named Camp Hoffman but was renamed after a 509th PIR trooper that had been killed in North Africa. Again, the men were housed in tar paper barracks, but unlike Toccoa, these single-story buildings were complete, with wood floors, steel-framed beds with springs, mattresses and sheets, and screened windows. They even had heat if needed. The barracks were broken down into company areas, with enough buildings to not only house the troops, but the Company Headquarters, supply section, and mess hall. The dirt track streets on which the buildings were located were lined with cut pine logs to differentiate the walkways from the streets. There were even street signs with names like, "Currahee Road", and in honor of their battalion Commander, LTC Robert Strayer, "Stray Wolf Alley." One look around and it didn't take long for the men to realize they were in the land of sandy soil and pine needles.

Left: Dog Company trooper Don King stands amid the street signs at Camp Mackall. Note the signs for "Currahee Road" and "Stray Wolf Alley," in the Second Battalion Area. (Photo courtesy of the RD Moellenkamp Collection via Susan Moellenkamp Saracino) Right: William "Bill" Guarnere and James "Punchy" Diel strike a pose outside one of the Second Platoon barracks buildings at Camp Mackall. The two had become close friends during their time in the States. Already suffering from the loss of his brother Henry when "Punchy" was killed in Holland, Bill said, (Punchy's death), it knocked me for a loop." (Photo courtesy of Amos Taylor)

Carwood Lipton remembered when they were sent to Camp

Mackall, NC, "Where we joined several other parachute regiments for continued training in all types of missions and in strenuous, physical training and parachute jumps to prepare us for airborne combat operations."[201] Pat Christenson was not impressed with Mackall at all. "The chow was terrible. There were too many inspections, and we spent too much time pulling guard duty," which seemed to consist of collecting up the numerous pine cones that dotted the area around the barracks. They also found that their barracks were infested with bed bugs. So, the men set up their pup tents in the common areas of the Battalion area while the buildings were fumigated.[202]

There was a bit of a rude awakening on the first day in camp, when at 0545 hours, a cannon was fired to signify it was time for reveille. The men were not expecting that kind of wakeup. It also signified the start of their next phase of training. From early February all the way through March, the men were put through their paces and were inculcated to what their lives would be like while at Mackall. For the most part, it was much like their time at Benning after they completed jump school, and that meant more forced marches of varying lengths, weapons maintenance, field problems and training jumps.

On February 18th, the company took part in a field problem that began with a practice jump. They wore newly issued jump suits and carried their full load of equipment. "Gordy" Carson, jumping number two behind Lewis Nixon, recalled the jump being a clean one with no chute failures or injuries, and as a result considered it a "fun jump." The field problem was focused on squad maneuvers. Since it was only a one-day exercise, the men were back in their barracks that evening, with many of them looking forward to catching a movie and hitting the rack early.[203]

James "Punchy" Diel and Lewis "Blackbeard" Nixon look over a map during a field problem at Camp Mackall. By February of 1943, Nixon had been assigned from Easy to the Battalion S-3, but he always found time for his old company. Diel, in whom Winters always saw great potential, was quickly promoted up the ranks to platoon sergeant for Second Platoon. (Photo courtesy of the Richard Winters Collection, AHEC)

Despite the men's excitement over the improved facilities, the training routine quickly became repetitive, tedious, and boring in nature. On occasion, tempers flared, and fights broke out. "Moe" Alley and Frank "Cowboy" Grant had a history. Neither liked each other and no one could recall what set them off. But according to "Smokey" Gordon, it came to a boiling point at Mackall. "They fought in the barracks, outside the barracks, it went on for a protracted period of time and I thought they were going to kill each other. But Alley was not one to ever back off from anybody. (And) Cowboy Grant was half crazy." Eventually cooler heads prevailed, and the pair was separated. Grant would eventually be transferred out of Easy Company in April.

Of the training, "Bull" Randleman said, "We trained before, during and after missions to the point of lunacy. A good soldier should never have to think what to do. It comes automatically."[204] And that response came from repetition. From Monday to Wednesday of each week, the men would take part in calisthenics and organized athletics, pitting them against other companies in the battalion, and sometimes, the regiment. They also took classes on hand-to-hand combat, communications, demolitions, and trips to the firing range. But it was the latter part of the week where Easy would put all those lessons together. They would typically be told that they would be going to the exercise areas for two-, three-, or even four-day field problems. Sometimes the men would conduct a training jump with full equipment, in advance of the exercises.

Two photos comparing Easy troopers as ease at Camp Mackall. Knowing that they had completed basic training and had become jump-qualified lifted a huge weight off of the men. Sobel was still tyrannical, but they had made it past their first two tests. The men could relax a bit more, despite the heavy field training regimen.

The photo on the left features Skip Muck. Next to George Luz, Muck was probably the most well-liked member of the company. Seen here, he appears to be enjoying some down time with his buddies near the Easy barracks at Mackall. According to Don Malarkey, Skip radiated a warm personality that drew others to him like bees to honey. (Photo courtesy of the Muck Family via Eileen O'Hara)

The photo on the right depicts Dewitt Lowrey. A Toccoa Man, strong and determined to be a good soldier, Lowrey always pulled his weight. Here he relaxes with a cigarette on his mattress on the railing outside one of Second Platoon's barracks buildings. The men didn't have much in the way of creature comforts, so just sitting on a padded mattress was probably considered a luxury. (Photo courtesy of Robert Burr Smith via Susan Smith Finn)

Easy Company's training started in earnest on March 2nd, 1943. The training area was an hour's walk from the company area, a place that Winters referred to as "Piney Woods." The men began conducting daytime field problems and, in some cases, the weather did not cooperate. The temperatures would drop, and it became cold and damp. They even suffered through a snow flurry on March 4th. For men that had spent the last six months in the mostly warm southern part of the country, the sight and feel of snow was a definite oddity.

A few days later, Easy kicked off its night field problems which ran from 2000 hours to 0800 hours the following day. The first of these night exercises featured First Platoon in the defense and Second Platoon attacking them. It rained on them the whole time. "Skinny" Sisk managed to find a way to avoid the rain, but in the process, was almost caught red-handed in doing so. "I found a little shack and built a fire (and found) a (steel) frying pan. Sobel must have seen the smoke. He came by (the shack) and was peaking through the window. I threw a stick at (the window). (Most of the time on exercises) we didn't have the sense enough to come in out of the rain. It was pouring down raining. Somehow, he (Sobel) he either seen or recognized (someone was in the shack). (I) figured it was just another soldier, but it was Sobel himself."[205]

By the time the exercise ended the next day, the men, outside of "Skinny," were thoroughly soaked and thoroughly pissed off. Robert T. Smith spoke for many of the men when he wrote his parents about the field problem. "We left camp Thursday morning in the rain and that night it turned to sleet, the same thing happened Friday; and we marched home this morning likewise in the rain. We didn't have a dry minute the whole time. Quite miserable." The griping and bitching continued straight through the truck ride back, that is until they were told they were being given a three-day pass. The men began formulating plans before they had even returned to the barracks. "Gordy" Carson, "Albie" Blithe and George Luz, took a bus to Charlotte, had a wonderful time and returned the next day. Bob Mann went to Rockingham on pass and got stinking drunk on white lightning. The next day, he fell out for morning formation wearing a pair of roller skates and not even he could say where he had gotten them. Spotting the skates, First Sergeant Morehead just stood there, stunned, clearly never having seen anything like it in all his years of service.[206] The day after the "roller skates incident," it began raining and continued unabated for several days. As a result, training problems were suspended until March 16, 1943.

While at Mackall, one of the strangest episodes in the Regiment's short history finally ended. Prior to the epic march from Camp Toccoa to Atlanta, a sergeant by the name of Sherman Irish, convinced Sobel to let him drive his personal car to Atlanta, with all of the captain's belongings. Ed Tipper recalled the story as such, "Company E had a sergeant, Sherman Irish, sort of a golden boy who smiled all the time. He was a favorite of Captain Sobel's, or certainly didn't have a conflict with him like others did. Irish was competent; everybody loved the guy. Well, when the time for the big hike came along, Irish went to Sobel and said: "You have a car, I'd like to make the hike, but somebody needs to take your car to Benning, and I'd rather help you with your car." Sobel had a newer model Ford and agreed. Irish further volunteered to pack along any of the other officers' valuables in the car. Other officers obliged. The day of the hike came. Irish got in Sobel's car. The rest of us started walking. We arrived in Benning, but Sobel's car was nowhere to be seen. Sobel received a telegram saying the car's transmission had broken down in Atlanta and that Irish needed $100 to get it fixed. Sobel sent the money but the car still never arrived. Nor did Sherman Irish. He had sold all the officers' items including the wheels off Sobel's car."[207] The military police were informed, and Irish was finally declared, "Absent Without Leave", AWOL for short. He was officially on the lamb.

Of course, the men in the company used the incident to torture Sobel. After a while, fake updates on Irish's location began appearing on the company bulletin board. "Popeye" Wynn found that especially entertaining. But that wasn't the only thing done to stick it to Sobel. Ed Tipper recounted that, "Weeks afterward when we made our jumps and were on field problems, every lone car that appeared raising dust on a nearby country road brought shouts of, 'Here comes Irish!'" Suffice it to say, Sobel didn't find the taunts funny, and it just caused him to get even more petty with his punishments for whatever infractions fancied him.[208]

A full three months later, on March 10th, the whole episode finally ended. Irish was apprehended and placed in the 506th stockade to await court martial. When he was finally brought in front of the court, his defense lawyer put on a vigorous defense. Irish was eventually demoted, fined, and transferred out of Easy, but remained in the Regiment. This infuriated Sobel because every time he saw Irish he was reminded of what happened.

With the conclusion of the trial, Easy headed back out for a two-day field problem, but not before another round of transfers occurred. The entire cadre of original NCO's, Joe

Stedman, Norman Ford, and Steve Kudla, were transferred out, as was solid First Platoon noncom, Jim Benton. Benton would go on to serve with distinction with D Company, 517th Parachute Infantry, earning a Bronze Star and Purple Heart.[209] The others formed the cadre for another unit and went off to begin training those fledgling paratroopers. First Sergeant Morehead also transferred out to the 508th Parachute Infantry Regiment, but not wanting to leave the 506th, Morehead took a reduction in grade to staff sergeant and requested a transfer to Second Battalion's Fox Company.[210] Under normal circumstances, this would have been a crippling loss for any unit, but Easy now possessed a core of homegrown NCOs, men who had served from the earliest days of the unit's existence, men who were almost universally respected within the company ranks.

For the next couple of weeks, Easy continued training... firing range time, graded field problems, conditioning hikes, although at least two training jumps were canceled due to poor weather conditions. On May 23rd, the company moved to Camden, South Carolina, and conducted more field training until the 28th. To the men, the training continued to be repetitive and morale-sapping, that is except for a live fire training event that took place on the 26th. Robert T. Smith experienced a bit more excitement than his fellow First Platoon troopers. "This last week has been by far the toughest and I was really worn out when we got in yesterday afternoon. The last feature of the week was a full-scale battle with live ammunition; I guess they wanted to see how we could take it. It was exciting and really made you watch your p's and q's; one guy didn't and so he won't have to jump anymore, altogether there were only four casualties which is darn good for this kind of stuff. You see there were land mines planted and if you stepped on one it blew up about ten feet from you covering you with mud and dirt, also, they had dynamite in the trees and if you tripped its wire the whole treetop blew up and scared the dickens out of you. One of them went off over me and I jumped so high that when I hit the ground, I thought that I had just made another parachute jump!"[211]

Left: The men of Second Platoon go over maps in preparation for a two-day field problem at Camp Mackall. Sharp-eyed viewers will notice the men are wearing newly-issued, unreinforced M42 jump suits.
Standing, left to right: Unk, Skip Muck, Dick Winters, unk, Punchy Diel, unk, Cleveland "O'Neill" Petty, unk
Kneeling, left to right: unk, unk
(Author's Collection)
Right: From left to right, First Sergeant William Evans, Dick Winters and Punchy Diel lead Easy back from the two-day field problem.
(Author's Collection)

Another morale-sapping incident occurred around this time, when the company lost Clarence Hester, the original company executive officer. Lieutenant Colonel Strayer had seen Hester in action enough to know he was a talented officer with an eye for detail, so he transferred him up to Battalion Headquarters as the Assistant S3 Officer. It was another tough loss for Easy, as Hester had been somewhat successful in his attempts to shield the other officers from Sobel since the company had left Toccoa. The loss of Hester would have a trickle down effect on the other officers.

Finally, on 30 March, the weather cleared sufficiently to conduct a training jump. The men had only had six jumps to that point. The company headed out to the Parachute Packing and Supply Area near the Mackall Military Airfield to pick up their chutes at 1300 hours. The objective was for each platoon to land in their own separate area of the drop zone, then come together at the company assembly area. This was accomplished with little difficulty, and even less fanfare.

Camp Mackall was also where the men experimented on how to parachute with different weapons and loads, and the practice here would eventually benefit them when the time came to do it in combat. For "Smokey" Gordon, nothing terrified him more than the idea of having to jump with his .30 caliber light machine gun. "Well, I'm not going to do it," he told Platoon Leader Wally Moore. Moore shook his head. Finally, "Popeye" Wynn sidled up to Gordon. "Don't worry Gordon. I'll jump the gun for you." Wynn weighed about 140 pounds. Gordon became incensed, thinking of the much smaller and lighter Wynn jumping his gun. His pride wounded, Gordon growled back, "You little pygmy. I'll jump the gun." And he did.[212]

"Jumping equipment was a challenge and dangerous duty, for the equipment was not designed for parachute operations. One such example was the bazooka as it had to be jumped in one piece. Light machine guns had to be jumped in one piece, tripod separately. Likewise, the 60mm mortar and base plate. The M1 rifle was broken down into three pieces and placed in a Griswold bag or gun container and strapped on the parachute harness with snap fasteners. All food, ammunition, maps, hand grenades, and more were also attached to the body in some fashion and was believed to sustain you for three days. So, when you practiced jumping, especially at Camp Mackall, every jump was 'sweat' time!!" recalled First Platoon noncom Robert Rader.[213]

While at Camp Mackall, the men learned to pack aerial delivery containers. These would be strapped on the undercarriage of the C-47, and using a trigger system inside the aircraft, the jumpmaster would release the containers just before the stick jumped. Theoretically, the men would then retrieve the containers and utilize whatever supplies were in them.
Top: Skip Muck, Salty Harris and Paul Rogers inspect equipment laid out for inclusion in the aerial delivery container. (Photo courtesy of Amos Taylor)
Bottom: Once the inspection was complete, the canvas sides would be folded over the equipment, then the padded "sheet" would be rolled, forming the bundle, and then end caps would be affixed and parachute pack attached. This sheet displays the container for one of Easy's machine gun sections. (Author's Collection)

Carson was positioned behind Nixon, who was jumpmaster for the stick. Luz and "Duke" Dukeman were behind Carson. Everyone got out of the aircraft fine, but there was quite a bit of wind across the drop zone. Despite the soft, sandy soil, Carson hit hard and then was dragged across the DZ until he was able to collapse his chute with assistance from the others.[214] Joe Toye broke his leg and ankle. After landing, medic Al Mampre found himself scolding a trooper who was struggling to collapse his chute while being dragged across the drop zone. The trooper sheepishly looked up at Mampre as he stood over him. It was the Battalion Executive Officer, Major Oliver Horton. Mampre chuckled to himself. "Sorry, Sir."[215] Barney Cunningham, of Second Platoon, landed way off course, in a farm's chicken coop, breaking his leg. The family who owned the farm ran into a copse of pine trees when Barney first landed. Later that day, the farmer fed him fried chicken and Cunningham was sure it was one of the birds he landed on. He was stuck there for two days before the other men found him and evacuated him. That same evening, back at the assembly area, some of the men began NCO training classes.

Increasingly, by late-March, the company began taking part in night field problems, and it was because of this that the men of Easy got very comfortable operating in low- and no-light conditions. They could recognize each other by their silhouettes, the tilt of their helmeted head, the way they carried their weapon, or the way they walked. Many a trooper would just see the outline of another and know who he was. This calmness in operating in the dark would come in very handy when they finally went into combat.

This is not to say that everyone was comfortable operating at night. "Popeye" Wynn was Captain Sobel's runner for a short period. During one field problem at Mackall, "Popeye" 'purposely' got lost after a night jump, figuring he'd get a decent night's sleep while being free the tyrannical Sobel for the evening. He returned to the company the following day. While pleading his case to the obviously skeptical captain, "Popeye" explained that he couldn't see in the dark to which Sobel replied, "You need to learn to see in the dark." "Popeye" was perplexed as to how he should do so saying, "How the hell do I LEARN to see in the dark?!?!"[216] After being made to dig a 6x6x6 pit and fill it up, "Popeye" was relieved of his duties as runner and returned to Third Platoon, only to be replaced by Ed Tipper.

Ever since Toccoa, Tipper had harbored an intense hatred for his company commander. So, when he was presented with the opportunity to make Sobel's life miserable, he took it and ran. "With my help," Tipper recounted, "Sobel was able to mislay his maps, compass, and other items when he most needed them. He was getting similar 'assistance' from others and was disoriented and lost even more than usual. We were all hoping that he'd screw up so badly that he'd be replaced and we wouldn't have to go into combat under his command."[217] Another 'skill' that Tipper developed early one was his ability to scrounge food, even when the company was in the field. During one exercise, "Smokey" Gordon, "Moe" Alley and "Hayseed" Rogers noticed Tipper sneaking away. They later spotted him kneeling over his musette bag. "Hey Tip. Whatcha got there?", the three asked. Smiling, Tip said, "Nothing," closed the flap up on his pack and went back to the Company Headquarters. They later found out that he had managed to wrangle some eggs from a local farmer. Tipper, along with Johnny Martin (and later, Roy Cobb), would become one of the most proficient scroungers in the company.[218] [219] [220]

Members of Third Platoon relax outside their barracks at Camp Mackall.
Top row: Carwood Lipton, Ledley Pace, "Moe" Alley, Rod Strohl, Mike Ranney, Carl Fenstermaker, Dan West
Middle row: Ken Baldwin, Skinny Sisk
Bottom row: Bill Maynard
(Author's Collection)

One officer who was equally comfortable in the field and in the barracks was Dick Winters. Commensurate with his rank, Winters had been given the responsibilities of executive officer, but under Sobel, who loathed to reward anyone, he was expected to continue running First Platoon as well. This frustrated Winters as he felt he should have been assigned as XO permanently, while turning over the duties of platoon leader to his assistant, Second Lieutenant Archibald Barnwell. After Harvey Morehead's transfer, Myron "Mike" Ranney was appointed the interim first sergeant, that was until William "Bill" Evans was promoted permanently to that rank. Evans was a Sobel favorite, and in the opinion of many of the men, just as much a martinet.As a result, Winters felt like he was always putting out proverbial fires with no let up.

Sobel's inadequacies as a field commander continued to be laid bare at Mackall. At one point, during a night field problem, Sobel, hearing some leaves rustling in the wind, believed that the opposing forces were closing in on Easy's position. He quickly roused the officers and ordered them to deploy their platoons. After several hectic minutes of frenzied activity, everyone but Sobel realized the enemy had not arrived at the proverbial gates. Still, Sobel kept the entire company 100% awake through the entire night.

While Sobel could be hopeless at times, he understood the basic tenants of soldiering, including the need to keep your weapon within arm's reach at all times. Unfortunately, his efforts at reinforcing soldierly standards oftentimes backfired. This was the case the evening following the 'enemy alert' debacle. Sobel and Evans decided that they would teach the men a lesson. They would wait until the company had knocked off for the night, and as the men slept without guard, the two would slip into the perimeter, and make off with their rifles. Their plan came off without a hitch, netting the pair forty-five unsecured rifles. The next morning, Sobel began tearing into the men for their lack of discipline but failing to note that every one of them possessed their rifles. As Sobel continued his tirade, Captain Tom Mulvey, the commanding officer of Fox Company, along with several of his men, walked up to Easy's school circle, demanding the return of Fox's weapons. Sobel and Evans had gotten lost in the dark, wandered into Fox's perimeter and stole their weapons. Sobel sheepishly pointed to the pile of weapons, and the Fox men filed by to gather them. As they did so, most of Easy could barely contain their snickers and in some cases, outright laughter.[221] As a result of the embarrassing screw-up and resultant laughter at his expense, Sobel worked the company ragged for the remainder of the field problem. The men were surly afterwards, wearily stumbling back to their barracks the next day, cursing Sobel's continued harassment.

Left: Troopers from Easy's Second Platoon mug for a photo in the Company Area at Camp Mackall. It was unlikely that Captain Sobel was in the area when this was taken. Wearing their overseas caps in such and unauthorized manner would have garnered each of them a session digging a 6x6x6 pit. From left to right: Joe Toye, John Sheehy, Don Malarkey, Bill Guarnere, Skip Muck and J.B. Stokes. (Photo courtesy of the Guarnere Family) Right: Robert Burr Smith clowns with an M1A1 Bazooka in the company area at Mackall. The building to the rear of the photo is the company mess, based on the presence of the unit's cooking pots air-drying behind the building. (Photo courtesy of Robert Burr Smith via Susan Smith Finn)

Another time, Sobel ordered Platoon Leader George Lavenson and noncom Robert Rader to go dig a series of foxholes on a Saturday night. But the punishment backfired, as both were exposed to poison ivy and had to spend several days at the base hospital being treated. As a result, Sobel lost a platoon leader and squad leader for several days during a hectic training cycle. The worst part of the episode was that Rader had no idea what the pair had done to warrant the punishment.[222]

Sobel continued to be the butt of jokes. At one point, medics Al Mampre and Ed Pepping decided to have a little fun at the expense of two of the Battalion's officers. During one field problem, the umpires had designated Sobel and D Company Commander Captain Jerre Gross as mock casualties. Mampre and Pepping had assembled a casualty triage tent where they could work shielded from the elements. The two medics hatched a plan and quickly went to work. Mampre and Pepping informed the two captains that they would be injecting them with saline water to simulate morphine. What Gross and Sobel didn't know was that Mampre and Pepping had shot them up with sodium pentothal. The mild anesthetic soon knocked the pair out. And that's when the medics proved that an idle mind is indeed the devil's playground.

Mampre, in a truly diabolically funny maneuver, made an incision on Sobel's abdomen to simulate a cut for an appendectomy. He then sutured and taped over the wound. As Mampre sewed his 'patient' up, Pepping proceeded to shave off half of Gross' prized mustache. You see, Gross was one of the few men in the 506th to be allowed to wear a mustache. It was a point of pride to the captain and with his wedding a week away, Gross would be horrified over the missing facial hair. With the shave complete, Pepping then administered the coup de grace by putting Gross' right arm in a plaster cast. The "Band-Aid Bandits" had struck again!

Sobel, infuriated, tried to have the medics brought up on charges. But since they belonged to the Battalion's Medical

Detachment, they weren't under his command and as a result, both medics avoided charges. While Gross was mad about the loss of half of his mustache, he dutifully shaved the remainder off in preparation for his wedding. The harder part was explaining what happened to his soon-to-be-bride, who ended up very pleased as she was no fan of his facial hair.[223] [224] [225]

As March slowly slid into April, the men began to see a slackening in the training. Organized sports became more common, with the boys of Easy beating both Fox and Dog in spirited softball games. Money changed hands quickly as each company's gamblers looked to make a score while they sat alongside the ball field. Once again, Johnny Martin made book as well as a reputation as a first-class "operator."

Members of Easy's Third Platoon pose in their newly issued M42 jump suits in the company area at Camp Mackall.
Rear: James "Moe" Alley
Middle row, left to right: Terrence "Salty" Harris, Carl "Dutch" Fenstermaker, Paul "Hayseed" Rogers, Kenneth Baldwin, Carwood "Lip" Lipton, Cecil "Park" Pace, Forrest "Goodie" Guth
Botton row, left to right: Clarence "Crash" Tridle, Elmer Schuyler, Wayne "Skinny" Sisk, Lavon Reese, Darrell "Shifty" Powers
(Author's Collection)

Yet, Sobel continued to harass his young charges. John Connell, who had witnessed several troopers being injured on the obstacle course back in Toccoa, was a strapping Third Platoon man from Pennsylvania. At 24, he was a bit older than most of the other Easy troopers and as a result, was a bit wiser to how the world worked than his fellow soldiers. Connell had raised the ire of Sobel by returning a few minutes late from pass. Sobel called him into his office and began to berate Connell over his tardiness. As he received his tongue-lashing, something began to build in Connell. He began to seethe as months and months of frustration began to bubble to the surface. Finally, having enough of Sobel's over the top accusations, Connell began firing back. The two argued back and forth for several minutes. Sobel wound up his tirade with, "I expect the men in this company to do right and learn to be good soldiers and learn to obey orders because I might have to depend on you in combat." Connell seized the moment to get the last bite of the sandwich. "You'll never get no help from me...Sir." Speechless over the insubordination, Sobel grabbed an expensive pen set from his desk, threw it on the floor, and began stomping it into pieces. "Connell, you made me break my pen!" he screamed. Within days of the incident, Connell was transferred out of Easy Company. Connell had cleared his mind and conscience, but it was Sobel that had the last laugh.[226]

In mid-April, the men were put through some basic jujitsu and judo training. "Shifty" Powers recalled the first Judo and wrestling class. Dick Winters was a man "Shifty" said the men of Easy Company trusted readily. "He was a hero at what he did – a real good officer and a fine tactician. I remember one of the first times he stood in front of us and asked if anybody wanted to wrestle him. Nobody responded. 'Any two of you like to wrestle me?' Nobody. 'Any three of you like to wrestle me?' Well, three fools stepped forward, and he beat the shit out of them. Winters was always in top shape. He didn't drink or smoke. He was what you thought an officer should be, he took responsibility for what had to be done. He'd be friendly with you, but it had to be one on one."[227] Winters' reputation as a wrestler, which began in Toccoa, stayed with him, and no one dared to ask him to wrestle again, until just before D-Day.

Soon after, it was back to the field. On 19 April, Easy began a two-day field problem until, during which time the men executed a 25-mile hike, which took them seven and a half hours to complete. They had two hours to rest afterward and then moved out to the maneuver area, where it began raining right after their arrival. They set up their bivouac in the pouring rain, and then ate a late, soggy supper. After a brief rest, they started the field problem at 0200 hours, maneuvering and battling until noon, before finally being trucked back to Mackall by 1830 hours. The men, with pruned skin, and wet, sand-encrusted uniforms and boots, huddled together in the back of the deuce-and-a-halves for warmth. It had been a trying, tiring couple of days for Easy, but they would soon be back at the barracks, humming along to Glenn Miller's "Little Brown Jug," or headed off to the base theater to see, "The Moon is Down", featuring Cedric Hardwicke.

Soaking wet from the rain, Third Platoon troopers attempt to relax in the platoon bivouac area during a field problem at Camp Mackall. Pup tents are clearly visible in the right rear of the photo as are the Thompson submachine guns carried by Talbert and Wright.
Top row, left to right: Floyd "Bunny Duck" Talbert, Lavon Reese, Paul "Hayseed" Rogers, Rod Strohl, Richard "Red" Wright
Bottom row, left to right: Terrence "Salty" Harris, unknown helmeted trooper with back to the camera (likely a medic based on the harness he wears)
(Author's Collection)

The last days of April would be some of the busiest that the company would experience while stateside. After qualifying with their individual weapons on the rifle range, Easy was ordered to take part in an Army-wide physical fitness test. They ended up scoring the highest company average grade for the entire 506th. Battalion Commander Strayer recalled of the test, "I remember at Camp Mackall that the Department of the Army had (us take) a battalion physical fitness test that lasted several hours. Second Battalion scored something like 97% which was the highest ever recorded for a battalion and they made us do it all over and (the second time we) had all cooks and service personnel involved. We scored over 98%. The officer from Washington, a Colonel Jablonski, who was checking these tests, told me that we scored higher than any battalion in the army! Sink just wouldn't believe it. Just goes to show you what a great group of officers and men we had."[228]

Many of the boys thought that after scoring so well in the fitness test, there would be some reward. There was...in the form of a very detailed company-wide inspection. Sobel toured the entire company area to include the mess hall, and then had the men fall out by platoon with every piece of issued gear for him to personally inspect. The men seethed the entire time, feeling like Sobel was treating them as if they were back at Toccoa. To add further insult, once the inspection was complete, Sobel had the men take part in a company parade, in full Class A uniform. Finally, at the conclusion of the parade, as the sun began to dip in the sky, Sobel issued passes out to the men. Good for 24 hours, some went into town and got drunk, while others headed to the movies to see "Slightly Dangerous", starring Lana Turner. The theater echoed with catcalls as the beautiful blonde moved across the screen, helping the men to mentally put aside all that they had endured to that point.

As much as the drink and movie helped the men to forget, they were back at it before they knew it. An infiltration course punctuated another four-day field problem, where they had to crawl a hundred yards while machine guns were fired over their heads and TNT was blown up around them to simulate artillery fire. The course had a small lake nearby and at the conclusion, the men were permitted to go swimming before heading back to Mackall. A short-in-stature, but wiry Irish-American lieutenant stood with the other officers, observing the men going through course. Harry Welsh, newly commissioned while with the 82nd Airborne Division, had been assigned to Easy's First Platoon as the assistant platoon leader. Time would eventually prove that the 82nd's loss would be Easy's gain.

May began much like April ended. Range time, communication classes, and another fitness test, this one sponsored by the Airborne Command, filled the early days of the month. Easy also gained more troopers from transfers with Dick Davenport (Second Platoon), John Lee Eubanks (Third Platoon), Joseph "Red" Hogan (First Platoon), Eddie Joint (Second Platoon), Lewis "Bob" Lampos (Second Platoon), and Robert "Jeeter' Leonard (First Platoon), all joining the company at the midpoint of the month.

Much like Lipton, Joint had read about the paratroops and found the idea appealing. But the realities of the job set in quickly with him. "I joined the army while everybody did it and I think that I chose the Airborne because I would like to jump out of airplanes...that was not too smart. Nobody told me that they were going to shoot at me." If he had any reservations, his new comrades-in-arms didn't notice them.[229] The new men were quickly incorporated into their platoons and were measured by the others over the course of two training jumps with their new unit. Each of these six new men would finally be accepted by the others and eventually make the D-Day jump with Easy Company.

A mix of First and Second Platoons, Easy Company pose at Camp Mackall.
Front row, left to right: Dewitt Lowery, Chuck Grant, Barney Cunningham, Rod Bain, and Joe Toye. Back row: Burr Smith, Warren "Skip" Muck, Don Malarkey, Denver "Bull" Randleman, John Serilla, John Sheehy, Tom Burgess
(Photo courtesy of Don Malarkey)

May the 7th saw the Easy deploy for a two-day field problem. At first the men thought they'd be getting a break, as they were driven halfway to the exercise area. Then they had to march the last five miles before kicking the field problem off, but since the march took place in the middle of the day with sun high in the sky, the men were quickly sapped of their energy. Sleep was tough to come by that night and the next day, they formed up to go fire on one of the base's many combat ranges. Easy had an observer as they fired, General Elbridge

Chapman, Commanding General of the 13th Airborne Division. The company gathered around Chapman after the range went cold. He had only one message for them. "Gentlemen, well done today. You should know, your time here in the U.S. is limited." And with that, they broke ranks and climbed on the trucks for the ride back to Mackall.[230]

On the 17th and 21st, Easy conducted two more jumps, the eighth and ninth for the company. On the 17th, "Gordy" Carson recalled jumping eleventh in the stick, sandwiched between Coburn "Johnny" Johnson and George Luz. The jump went smoothly and thankfully, the Easy men incurred no injuries, except for Sobel, who injured his ankles. For the rest of the company, they were off on another two-day problem, with the main objective of attacking and seizing the small town of Hamlet, North Carolina. The company jumped clean, and after assembling, moved out to where they would conduct the attack. After they dug in, they went to 50% security and tried to get some sleep. The attack was scheduled for dawn, and as the sun started to climb into the sky, Easy came boiling down the hill above Hamlet. They quickly captured the town. After pulling back to a spot near a small pond, the company dug in again to defend what they had captured. No attack materialized and once again, they boarded the trucks for the ride back to Mackall.

For the next jump on the 21st, Carson was the number 12 jumper in a stick jumpmastered by Mike Ranney, who was now the platoon sergeant for First Platoon. With his injury on the previous jump, Sobel remained at Mackall with Evans, turning the company over to Winters for the ninth jump. The jump and field problem went well, but the company had to march the eight miles back to camp in the heat and did so sweating and swearing the whole way. To alleviate the stress, "Bull" Randleman called "Skinny" Sisk out of the formation to call a cadence. "Skinny" began belting out "Life is Like a Mountain Railroad" and soon everyone was happily singing along with him.[231] Unfortunately, the swearing picked back up after they got back as Easy was directed to the sheds to start packing chutes for the next jump.[232]

Left: After receiving their chutes at the Parachute Packing and Supply Area, Easy Company begins the walk to Mackall Military Airfield for a training jump. (Author's Collection)
Right: Third Platoon trooper, Jerry Wentzel, stands fully kitted out for a training jump at Camp Mackall. His entrenching tool is taped to his leg in order to avoid becoming tangled in his shroud lines after exiting the aircraft. (Author's Collection)

While the men had been out training, Sobel and Evans decided to inspect the barracks for infractions. Sobel and Evans went through the barracks with a fine-tooth comb. Nothing and no one were spared from the inspection. Their footlockers were opened and the men's personal items rifled through. Sobel was infuriated. He found a litany of what he considered contraband. When the men arrived back at the barracks, they found all their worldly possessions stacked up on their bunk mattresses...letters, civilian clothing, hygiene gear, underwear (called skivvies by the troops), pornography, and ammunition. "Gordy" Carson recalled, "We walked into our billets. Here everything we owned was piled on a big pile on our bunk (sic)...everything we own, underwear, toothpaste. Everything was piled right on top of all our bunks. I don't know what the hell he was looking (for)."[233]

Sobel had posted a list of all the confiscated items and whose footlocker they had been found in. The men seethed and cursed Sobel but none was more incensed than Ed Tipper. "Everything we thought of as personal property was in disarray, many items missing...practically every soldier had something confiscated. From me he got three number ten cans of fruit cocktail and sliced peaches which I had stolen while on K.P. and planned to share with my buddies, and a couple of blocks of TNT taken from a demolitions class...some individuals lost expensive, officer-style shirts, which were never returned. One soldier had been collecting prophylactic kits...a few condoms were acceptable evidently, but 150 or 200 constituted contraband." But what really set "Tip" off was that he had met a young lady from Rockingham, North Carolina on pass several weeks back. The two had maintained a vigorous correspondence since that date. Now he found all the letters, some personal in nature, opened and scattered across his bunk. What had been hatred for his captain, was now incendiary.[234] [235]

Clustered around the Company Headquarters' bulletin board, men from each of Easy's platoon are present. Left to right: "Popeye" Wynn, Ed Pepping, Skip Muck, Joe Toye, Tom Burgess and Don Malarkey. (Author's Collection)

Finally, beginning on the 23rd of May, the men were rescued from the tedium of Camp Mackall and the chickenshit 'leadership' of Sobel. Colonel Sink had declared that all regimental, battalion, company, and platoon officers would be war casualties, and all non-commissioned officers had to lead a planeload of troopers on the next jump. Before heading to the airfield, the platoons were kept busy loading bullets into cloth machine gun belts. After that was completed, Easy was driven to Maxton Field in South Carolina and bivouacked near the airfield. Over the next few days, they went through several classes and practiced tactics and map reading. On the 27th, the men chuted up, and jumped into Kershaw, SC.

After landing, Easy assembled and marched 12 miles away, where they set up a defensive position. In "Rook" Rader's mind, Colonel Sink's "Plot worked...He proved any good non-com could do the job."[236] The next day, the weary troopers marched to Kershaw's airport, where they grabbed their chutes, suited up and flew back to Mackall. Carson recalled being situated between Johnny Johnson and George Luz again for the jump. It was their tenth jump and many of the men felt a sense of accomplishment at making it to double figures. There was a lot of hedgehopping on the flight in. The red light came on early, eleven minutes from the drop. The men, with harness straps tightened to the point of inflicting pain, just had to stand there and take it as the aircraft made their approach.[237]

Not every man's jump went smoothly. As Second Platoon's Don Malarkey exited the aircraft, he began scanning the sky to see where the other men in his stick were coming down. As he did so, he realized he was descending at a more rapid rate than the others. He quickly looked up and figured out the problem. He had four or five blown panels in his canopy, which was allowing the air to pass through the chute instead of filling it with air, which would in turn slow his descent. As he oscillated, Malarkey committed another parachutist's sin. He pulled the rip cord on his reserve, but instead of deploying the chute (by hand) in the opposite direction of his turns, he did so in the same direction. This caused the reserve chute to billow back into his face. Not only was he falling too fast, but he had a face full of parachute silk and couldn't judge the distance to the ground to prepare himself for landing. Before he knew it, he came crashing down through the branches of a tree near the edge of the drop zone, coming to a halt a few feet off the ground. After releasing himself from his harness and collecting up his chute, Malarkey realized that he had accidentally packed a shot bag in his canopy. The heavy, lead BB-filled bag had split apart when the chute deployed. The pellets acted like a shotgun blast, blowing through the panels. Filling it away as luck-of-the-Irish, Don smiled as he figured out that this was his tenth jump, meaning he would no longer have to pack his own chute.[238] [239]

Carson managed to land smack dab in a tree, thirty feet off the ground. After working his way out of his harness and down the tree, he managed to link up with Luz and Perconte. They hadn't made it far when the trio stumbled into a quick firefight. Backing away from the enemy position, they hatched a plan to take their time getting back to the company. "Why rush back," Carson thought. With Luz in the lead, they stumbled upon a small shack and knocked on the door. Perconte, with his thousand-watt smile, was able to convince the inhabitant, an older lady, to whip them up some hot food, which they gladly paid for. With full bellies, the three First Platoon men lounged around in the sun for a bit before finally getting back to Easy's lines.

After the field problem concluded, the men found out that for once, they would be flying back to Mackall and landing instead of jumping. For many, this was the first time they had ever taken off and landed in an aircraft. Most had only gone up and jumped. It was a new, odd feeling for sure. The flight home was smooth, and the men joked and laughed in the rumbling C-47's as the envisioned a weekend of passes, beer and girls. As they marched back into the barracks area, they found another new trooper waiting for them, Roy Pickel. The affable Pickel was quickly sent to the company office to get processed into his new home in Easy's First Platoon.[240]

9

The Graduation Exercise

"We were tested on maneuvers in Kentucky and Tennessee in June (and July) and made two tactical combat jumps there as part of the Blue Army."
- Carwood Lipton

Tennessee Maneuvers, 04 June – 21 July 1943

In the fall of 1942, the War Department decided that the U.S. Army would conduct a series of field maneuvers in Tennessee. The reasoning behind the decision to use Tennessee was due to two factors; the first was that there was a large area of sparsely populated land available and secondly, the land in question resembled France, Belgium, and Germany with its rolling hills, streams and rivers, and forested tracks.[241] Lastly, the area was chosen because of its proximity to railroads and federal highways which would facilitate the movement of troops and equipment into the maneuver area. Units would be assigned to take part in these exercises which were designed to simulate combat conditions. Red and Blue "armies" would face off against each other in a series of complex and realistic training exercises.

In early June 1943, the entire 506th Regiment was ordered to support these exercises not only in the state of Tennessee, but in Kentucky as well. These would come to be known to the men as the Tennessee Maneuvers. For Easy Company, they would be dropped behind the lines to establish roadblocks, destroy bridges, and disrupt communications. Standard fare for paratroopers. The exercises were divided into several phases, and Easy would take part in two of them, "Operation 7" and "Operation 8", spanning much of June and July of 1943. If Camp Mackall had been Easy's finishing school, then the Tennessee Maneuvers would be their graduation exercise.

According to Tennessee historian Frank Burns, "Lebanon was chosen as headquarters and Nashville as the principal rail head. Over the hills and valleys of twenty-one counties Blue and Red armies engaged in weekly strategic "problems," with troops moved in and out according to a calendar of "phases" that lasted about four weeks apiece. In the military's scenario Nashville was Cherbourg, without the bombing. The first and second problems usually took place east of Davidson County, but the third in each phase would pose attacking Blue troops against Red troops in defense around Donelson in Davidson County and Couchville in Wilson County. This force would advance to the east toward hilly terrain. In one instance, at least a (single field) problem involved the defense of Berry Field in Nashville against Blue airborne troops."[242] Easy was assigned to the Blue Army for the duration of the maneuvers, which would last from 4 June until 12 July 1943.

While no one was thinking "Normandy" at that time, the reality was that the names on the map mattered little. The Easy men just weren't focused on that. They had their own little slice of the pie to worry about. The colonels and generals could worry about big lines and arrows on the maps. For the men at the bayonet point, they only cared where they were supposed to land and what they were supposed to do once they did. In football vernacular, they would take care of the blocking and tackling. The prevailing attitude was, "Let Colonel Sink and Major Strayer worry about the big picture. That's what they get paid to do."

* * *

On the 4th, Easy Company boarded the train with the rest of Second Battalion, and headed to Sturgis, Kentucky. The trip took several days and after a very tiring ride, the men, smelling a bit rough due to the lack of showers, finally arrived on the 6th of June. After disembarking, they marched three miles southeast of the town and set up their bivouac. Pup tents littered the fields, and the men were handed K rations to eat. This would presage the future quality of meals throughout the exercises. K rations would be eaten in the morning, and bag lunches containing a bologna sandwich and a piece of fruit would be eaten for lunch. If there was time, the company mess would serve hot meals for supper. If not, it was K rations again. Regardless, the men believed they'd have grumbling stomachs for the duration of their time in Tennessee.

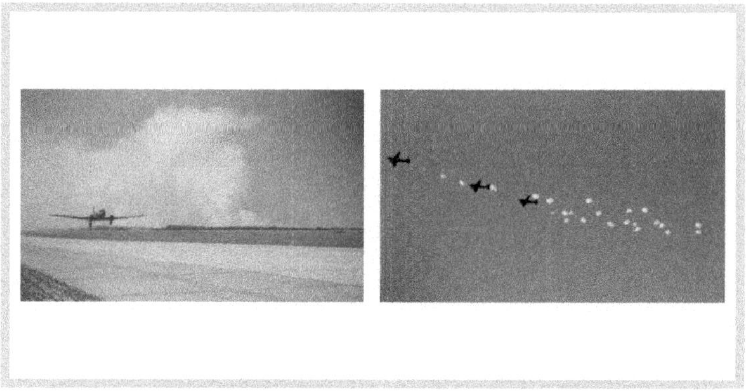

Taken by Amos "Buck" Taylor, these photos depict Easy's departure and for the Tennessee Maneuvers and their first jump during the exercise.
(Photos courtesy of Amos Taylor)

Grumbling stomachs was something the troopers would never get used to having. For twenty-something year-old men who lived with strenuous activity as their constant companion, the need for caloric intake was high. Simply put, they burned more calories than they ate. K rations did little to abate that problem. The K ration was a slim, boxed meal containing a tin of meat, usually pork, along with biscuits (think crackers), instant coffee with powdered creamer and sugar, instant drink mix, a fruit bar, chewing gum, cigarettes, and a small wooden spoon. The outer box was made of cardboard, and the inner box was paraffin treated cardboard, which in a pinch could be burned to use for heating the ration tin. Good in theory, but hardly practical in a combat situation. The Germans would later marvel at the K rations, but the men of Easy largely hated them after consuming them for weeks on end.

The maneuver area was very rough and sparsely populated, and the locals that lived in that area had a very difficult existence. Massachusetts-native Mitchell J. Dabrowski, a private first class in I Company, 329th Infantry who took part in the same exercise as the 506th as part of the Red Army, had this to say, "These maneuvers are pretty tough. In fact, it's about the toughest thing I ever had in the Army. Yesterday we were camping in some woods and got an idea to go to one of the farmhouses and ask them if they could fry us some chickens. The lady said she would. We told her to fry six. We came back at night and had the swellest (sic) feed I've had in a long time. Fried chicken, hot biscuits, milk, and raspberry pie. The whole works cost us $8.00 but it was sure worth it. If we ever come back, we are going to have her roast us some ducks. The way they live in the shacks around here is a crime. They are nothing but rough boards with clay pasted between the boards. I wouldn't live here for anything. But the people here seem to be very accommodating."[243] Rough was what it was and that's just what the Army wanted for its fledgling paratrooper force.

Dabrowski's sentiment was echoed by many of the Easy troopers who had interactions with the locals. Very nice folks, but their lives were, in a word, rustic at best. Many of the Easy boys spoke about the many dilapidated wooden structures throughout the exercise area, and many times they had families living in them despite the deplorable conditions. The locals were told by officials to avoid contact with the men and to do nothing to support them. Yet, many found time to cook up eggs or fried chicken for them which the boys gladly repaid with money or K rations or a combination of both. Either way, freshly cooked food that wasn't prepared by

Army cooks was always preferred.

On June 8th, 1943 "Operation 7" officially kicked off and with it, Easy Company's official participation in the Tennessee Maneuvers.[244] Reveille sounded early that day, to the tune of 0400 hours, and by 0500, they were assembled and on the march. Much like their time at Mackall, they trudged over to the airfield, only this time it was Sturgis Army Airfield instead of Lawson Field. With 10 jumps already under their belts, the men were no longer required to pack their own chutes. The riggers would take care of that for them.

After gathering near the hangar, Captain Sobel briefed them on the mission. They would jump into their drop zone, form up, send out patrols to find the enemy, and then dig in and defend the DZ. The Second Platoon also found out that they would have to live without the services of Bill Guarnere for a bit, as he had been tasked to perform umpire duties for the duration of the exercise. Finally chuted up, the men climbed aboard the C-47s. The propellers began turning at 1120 hours and soon they were rumbling down the runway. The mission that day would be to jump 10 miles east of Gallatin, Tennessee.

Low-level flights like the one shown in the photo on the left caused numerous airsickness cases across the company. The second photo shows how rustic the drop zones were that were used during the exercise. (Left photo courtesy of James "Pee Wee" Martin via Doug Barber, right photo author's collection via the U.S. Army Signal Corps)

The flight was made at a lower-than-normal level, and the warm air buffeted the aircraft, making for a memorable 150-mile ride. With the aircraft bouncing around crazily due to the hot air, it didn't take long for the men to start getting sick. Robert T. Smith, who had recently been promoted to sergeant and placed in charge of the First Platoon's Mortar Squad, remembered the flight well and its effect on his fellow troopers. "We jumped into the battle last Tuesday morning and it was some jump, we had quite a plane ride before we reached out objective in Tennessee and hence six of these stout-hearted steel-gutted men got air sick (I was not one of them) and so the bucket was being passed quite frequently and finally it was having so much use that we had to bring in another bucket to take care of the business."[245]

For Pat Christenson, this jump was memorable, but for reasons other than a smooth landing. The men on Christenson's stick were suffering just as much as those on Sergeant Smith's. It was hot and oppressive in the C-47 and the warm mountain air created turbulence that shook the aircraft violently. Denver Randleman, last in line of his stick, was the furthest from the door. The heavy turbulence sickened him quickly, so "Bull" took off his helmet, bent over and puked in it. The man standing next to him caught a whiff of the smell and then he too threw up. This continued to the door of the aircraft, and once the helmets had been emptied on the floor, the whole aircraft stank of vomit.[246]

Finally, at 1340 hours, Easy began jumping into the sky with their white chutes blossoming out above them. It was a clean jump, in name only, and with no issues landing, the men quickly formed up on the drop zone. Christenson had not thrown up but was holding everything in until the green light came on, and when it did, he launched himself out of the aircraft. The fresh air flowed over him and Pat's stomach woes thankfully subsided.

Don Malarkey remembered that first jump in some detail. "We would march from Sturgis to the airstrip and as we did that, they'd have the parachutes waiting for us. The chutes had your name on them and they'd pass them out to us. Then we'd play 'boy scout' games with a Blue Army and a Red Army. This gave the officers a chance to control larger units in the field."[247]

Stanley Horn, a correspondent for *The Tennessean* covered the initial Blue Army parachute drops and wrote extensively about the initial actions, "Out of the low-hanging mist west of Gallatin, swift transport planes of the Blue army dumped paratroopers in the rear of retreating Red troops at dawn this morning...following up the dawn attack with another west of Gallatin shortly after noon, the Blue forces by nightfall had two battalions well-established and holding their own in the midst of the main body of Red forces."[248] What Horn didn't witness was Easy's arrival. According to Winters, after landing, the company had to conduct a march to the perimeter which included a crossing of the swiftly running Cumberland River in rubber boats.

Once they arrived at the Battalion's perimeter, the men went about their duties of setting up the defense. One plan devised to assist the defenders was to use some explosives to knock down several trees. Then the men would form up into working parties and drag the felled trees across roads and paths leading to the DZ. It was hoped that the well-placed logs would slow down any enemy vehicles approaching their positions. Joe Toye, who had gone through demolition training while Easy was at Camp Mackall, had some specific tasks to accomplish to help with the defense of the drop zone. "They set up a problem, and I was on this demolition crew. I wasn't with my outfit.And we used to go out ahead of them and blow trees over the roads, you know, just obstacles to make them (the Red Army) work and so forth."[249] Toye, along with Easy's other demolitionists, Burr Smith, Rod Bain, and "Rip" Van Klinken, surely earned their pay that day.

After the paratroops landed and spread out in the defense, the second wave of troop carrying and supply aircraft came in with their bundles of ammunition, bazooka rockets, communications equipment, medical supplies, and rations. Reporter Stanley Horn observed this drop as well. "The line of transports wheeled slowly to the right and suddenly from the leading ship blossomed the white canopy of parachutes as the men began to jump. Intermingled with the white, green, and red parachutes denoted equipment dropped along with the troopers...soon over 800 lean, bronzed troops were deploying toward strategic roads and bridges."[250] [251]

This wasn't to say the exercise went smoothly. On June 9th, 1943, the day after Easy landed, the Battalion pushed out several patrols to make contact with the enemy. The Red army counterattacked towards the drop zone and overran part of Second Battalion. While the Battalion logbook and the Easy morning report only mention some troops being captured by Red forces, Horn covered the action for *The Tennessean* saying, "...Red forces closed in and surrounded the Blue troops. The Blues, gradually driven back from the roadblocks they had established east and west of Gallatin, leaving in their wake demolished bridges, were finally rounded up by the Reds."[252] Umpires soon called the exercise for the day, and the paratroops were trucked to Bowling Green Airport in Kentucky. Early the next morning, June 10th, the Easy troopers climbed on the C-47's and flew back to Sturgis Army Airfield. That same day, the 506th Parachute Infantry Regiment was officially assigned to the 101st Airborne Division. They would forever be known as Screaming Eagles.

* * *

Just because the Sink Men were back at Sturgis, didn't mean that they'd be taking it easy. The troopers continued to improve the camp and train...and dodge the numerous copperhead snakes that inhabited the area. Another jump was scheduled for June 12th and like the first jump of the maneuvers, despite the early morning hours, the air was thick and warm, and the low-altitude flight subjected the C-47's to increased turbulence. The effect on the jumpers was debilitating. So much so, Albert Blithe was so airsick that he couldn't even make the jump. "Gordy" Carson, standing next to a wobbling Blithe, nudged "RT" Smith to his front. "RT...look...Albie's about to lose his breakfast." Looking five different shades of green, Blithe soon fainted. Carson and Smith laid Blithe down and he ended up flying back to Sturgis with the air crew.[253] [254, 255] Sherman Irish also flew back. As Carson checked his harness and parachute pack, he noticed that Irish's chute was hanging out of his pack. He sat back down and watched over Blithe while the rest of the stick jumped.

Chutes fill the air above a drop zone during the Tennessee Maneuvers. (Photos courtesy of James "Pee Wee" Martin via Doug Barber)

It was a quick assembly for the rest of the company on the drop zone. Unfortunately, the enemy was there in force, including some halftracks. Easy was forced to withdraw due to the size of the enemy force. Easy was outnumbered and outgunned. They made their way to a secondary objective, a road junction, and dug in to defend the area. Not everyone had assembled on time. Forrest Guth managed to get lost after landing and took a while to link back up with the rest of the company, so he missed all the excitement with the enemy halftracks.

There was also an incident involving members of First Platoon, and this time it was Johnny Martin who was the ringleader. Martin recalled the incident in question. "Our mission was to become mobilized. You've got the Red Army and the Blue Army. So, when we hit (the) ground, we jumped down there. That was a night jump, I'm pretty sure.[256] And I got together with George Luz and Frank Perconte, they were both in my platoon. It just so happened that we got together and took off and we went up through these woods, and here was the Blue Army, camped in there. And I said, "Let's sneak in there and see what the hell's going on." We snuck in there, and there's a jeep sitting here...and we looked in the jeep, and the keys were in there and everything. I said, "Jump in, we're going to take this jeep." So, we got in there, started the jeep up in the woods, and took off through the woods, up through the mountains there, and they've got these old roads that had been washed out and all of that. Some guy that was the Major's orderly gets on a motorcycle, and he starts chasing us. We go down this road bouncing around. Frank Perconte flew up in the air, came down and hit himself on the side! And I was gunning the hell out of the thing. And pretty soon the motorcycle guy goes over the hill chasing us, and the road was a terrible road.

But anyway, we got out and we got down the road and then the next morning we found out that we had stolen the S2 Major of the Blue Army's jeep, with all these maps and a whole carton of rations in there. So, we kept that for a couple of days, and that major had the MPs looking for us, and we went into Nashville, and oh, we were all over the place! And finally, we were going down a road and the MPs were coming one way, and we turned around to go the other way and the MPs caught us there. They finally got us, and they took us back to Sobel. Sobel says, "I'm going to bust your ass, Martin!" I said, "Hell, that was the mission." And here's his colonel saying "You'd better bust him! You'd better bust him!" Sobel walked away and he came back and says (sic), "Don't worry about it." But anyway, it was quite a little experience."[257] Leave it to paratroopers to steal a jeep from their own headquarters, create all kinds of havoc, and then get off scot-free.

Once Martin, Perco and "Portagee" were cleared of charges, they rejoined the rest of Easy, two miles outside Sturgis where the company had bivouacked, post-jump. The rest of First Platoon had a great laugh over the whole incident, but most wondered why Sobel hadn't busted the trio. "Heck, maybe he's getting soft," one trooper mused. "Not likely," said another. The brief reunion completed, the fellas turned back to their pup tents and card games.

"Popeye" Wynn had reputation as a heavy sleeper and "Hayseed" Rogers decided to play a little prank on him to lighten the mood in the company's "tent city" at Sturgis. Rogers drew outpost duty one evening and it was that night he hatched his plan. As "Popeye" slept in his tent, Rogers knelt by the opening and called out, "GAS! PUT YOUR MASK ON!" Startled awake, "Popeye" grabbed his mask and donned it and quickly fell back asleep. The next morning Rogers gathered everyone around "Popeye's" tent and then woke him up. "Popeye" crawled out of the tent, still wearing his mask. The eyepieces were fogged over because he had slept in the mask all night. Rogers said, "He looks like a man from Mars!" The men that had gathered around howled with laughter at the sight.[258]

On the 14th of June, the Battalion hopped on trucks for the short ride to Sturgis. Once there, they emplaned and were flown down to Camp Campbell. Unlike today, Campbell was a very rugged camp, not entirely built out and ready to house large bodies of troops. It reminded many of the young troopers of Toccoa in its sparseness. The men were forced to sleep alongside the runway due to the lack of bunk space in the not-yet-completed barracks. Don Malarkey recalled Campbell as having a few barracks buildings and a post exchange, but "not much else."[259] The PX afforded the men a chance to pick up fresh razor blades, shaving soap and tooth powder, along with candy bars and sodas. The next day was an early reveille, with wake-up arriving at 0520 hours. After being roused from their fart sacks, the company moved to the camp's dirt airstrip where they were issued parachutes. Their objective would be to jump into a drop zone near the town of Lafayette, set up a defensive perimeter and protect the area from attack by the Red Army. By 0700 hours, chutes began blossoming over the DZ. Once consolidated, Easy and the rest of Second Battalion were firmly emplaced and spent the evening in the hastily dug holes that dotted the perimeter.

Easy had spent a bleary-eyed night manning their positions, expecting to be attacked. At fifty percent manning all night, little sleep was had. Plus, the men spent the humid evening battling chiggers and active imaginations. Slight noises and shadows played havoc with them as they peered into the brush, scanning for the Red Army troops. Daybreak brought out the sun and a quick breakfast. As the men tore into K ration boxes, word was passed to be on the lookout for the glider echelon. As the C-47's and the WACO gliders began circling the drop and landing zone, the Red Army, sensing an opportunity to catch the paratroopers at their weakest point, began their attack.

Left: Burr Smith mans a .30-caliber machine gun on Easy Company's perimeter during the Tennessee Maneuvers. Right: An M4 Sherman tank of the 20th Armored Division trundles past an abandoned farmhouse during the Tennessee Maneuvers. The paratroopers had a tough time dealing with the armored forces because of the inability to simulate anti-tank measures beyond simple roadblocks. (Author's Collection)

TIME Magazine sent reporter John Walker to cover the exercise, and he rode into the drop zone in a glider alongside

101st Airborne Division commander General William C. Lee. Walker described the descent into the landing area, "At 6:55 a.m. we were less than five minutes from our target—"Z-field," a meadow near Lafayette, Tenn., where the divisional command post was to be established. Ground mists had kept us fairly high. Ahead and below we could see other tow planes flitting along. Now gliders were cutting loose, swinging off to land in the little tree-bordered fields ahead...Around us, the demonstration was going full blast. The tow planes had swung back overhead and were dropping more paratroops. The sky seemed to blossom and fill with 'chutes, white for the men and bright colors for specific types of packed equipment. Several other wrecked gliders were lying on the field. Men were scurrying by, organizing their positions and setting up mortars, machine guns and bazooka positions. Enemy tanks and cavalry appeared, blasting away with blank ammunition in a confusion of swirling dust and bright sunshine."[260] [261]

Apparently, at one point on the 15th of June, General Lee decided that he needed to have an urgent conference with staff. There was only one problem for Colonel Sink...he was 150 miles away in Sturgis. After receiving the radio call that Lee needed him and the officers under his command, Sink gathered up his staff, donned parachutes, boarded a C-47 and jumped into "Z-field" thirty minutes later, landing almost on top of General Lee's headquarters tent.[262] Later that day, an enemy plane buzzed the drop zone. Ray Nagell, of B Battery, 321st Glider Artillery, who had been attached to the 506th for this portion of "Operation 8" witnessed the plane and its subsequent attack. "They had a little Cub that was flying around, and he had a bag of flour, and if he threw that flour down and it got close to you, then you were a casualty and you had to be taken to the hospital.

But it was just make-believe." Teenager Murray Miles, whose family farm sat in the exercise area, remembered this tactic as well, "There was a bridge crossing the Stones River right next to our property. Each Monday, the Army would 'bomb' the bridge. The way they bombed it was with a Piper Cub airplane and dropped a sack of flour on it. Then the army could not use the bridge for the rest of the week. They would place an umpire on either side to keep the army from coming across."[263]

Teenager Raymond Doyle Gaines of Lafayette, Tennessee, was in his early teens when the Tennessee Maneuvers began. He remembered the day the 82nd and the 101st Airborne Divisions landed in Macon County. "The entire community turned out to watch as the sky filled with parachutes and gliders, as they descended on the Sullivan farm, three miles west of Lafayette...These divisions [later] distinguished themselves in Sicily, behind the Normandy beachhead, in Holland, and, of course, in Bastogne, Belgium, at the Battle of the Bulge." Clearly, these exercises were of huge interest to the farmland folk in the exercise area. For people who lived a rather mundane life, seeing the Army, especially paratroopers, conducting maneuvers was a source of great excitement.[264]

After completing the defense of the drop zone, "Operation 8" of the maneuvers concluded for Easy Company and the Blue Army on the 16th of June. The men of the company were only too happy to hop onto the deuce-and-a-halves for the ride to Bowling Green Airfield. Shortly after their arrival, they boarded the C-47's for the short flight to Sturgis, where after landing, they were treated to a feast of (real not powdered) scrambled eggs and biscuits. They next day Easy was transported to Camp Breckenridge where they stayed for the next few weeks.

Beginning on June 28th, Easy spent several days qualifying with their individual and crew-served weapons. They also went to the range to throw grenades. For the most part, it was easy duty. They caught up with their sleep, conducted some more practice jumps and recovered from almost three straight weeks of exercises. Some needed recovery time more than others. Chiggers had plagued poor Don Malarkey all throughout the exercise, as if they had specifically targeted him. While at Breckinridge, he had to be admitted to the Battalion Aid Station as his back had become infected due to his incessant scratching. By trying to find some relief from their nagging biting, Malarkey had scratched his back raw and it had become infected. With a soothing balm applied by the medics, he could start the recovery process, but not before he was flown back to Fort Bragg for additional treatment.[265]

While Don recovered the remainder of Easy Company spent another week as part of the Blue Army. The men experienced an eventful July 4th, in that they spent it all over the back woods of Tennessee. "Gordy" Carson wrote about that day in comedic detail, "We got orders to move at 10 and we prepared to move out at 1200, but we didn't get off until late that afternoon...We shot crap(s) all afternoon, and I went down and up. If they could get us any further in the Tennessee hills I would like to know it boy, we are at the end of civilization right now. We ate a late supper out here, and then got in trucks to come up to our outpost. The night was a heller, we must have rode on every back road in the Tennessee hills. We bumped around all night trying to find our spot, but no luck. Finally some of the trucks got stuck and we had to walk back down the stinken (sic) mountain."[266]

Left: Paratroopers descend upon a farmer's cleared field during the exercise. (Author's Collection via the U.S. Army Signal Corps) Right: The sky is filled with chutes during one of the many parachute drops during the Tennessee Maneuvers. (Author's Collection)

Despite being motorized for the better part of two weeks, the exercise culminated with two additional parachute jumps in mid-July. The men enjoyed seeing their jump logbooks getting filled up with new entries. They also received a replacement trooper by the name of Robert Marsh, who was quickly assigned to First Platoon. He would later become a mainstay of both the platoon and the company, serving until the war's end. As the men performed housekeeping around the campsite, word was quickly passed that the Regiment would be heading back to Fort Bragg, and eventually, overseas. The men had become bored with war games and could feel themselves inching closer to the war. While the prospect of combat didn't excite any of them, most possessed a "let's get this over with" attitude.

Easy and the rest of Second Battalion performed well during the maneuvers. So well in fact, that General William Lee, the Division Commander, sent Lieutenant Colonel Strayer as note to express just how pleased he was with the Battalion's performance.

"Our maneuvers for the period June 5th to July 15th have just ended, and for you and your battalion it should be with a feeling of gratification for a hard job well done."

"Results of the recent Tennessee Maneuvers indicate that your organization made a splendid showing. This indicates a high standard of training and competent leadership on the part of the officers and enlisted men of your unit."

"I desire to commend you and through you the officers and men of your battalion for the splendid aggressive action, sound tactical doctrine, and obviously well trained individuals. Your unit is especially commended for their action on the night of July 14-15 when they captured a large detachment of the enemy and assisted in the river crossing which was so successfully carried out. Every effort should be exerted to maintain this excellent record and I am confident that future tests will reveal further indications of excellent training and leadership."

By the 21st of July, Easy received orders to proceed back to Fort Bragg, North Carolina in preparation for movement overseas. They had no idea where they were heading but they knew it would happen soon. Carwood Lipton summed up the training Easy had endured as the Tennessee Maneuvers concluded, "It built up your physical stamina, our confidence in our own resources and in our unit, and our ability to withstand adverse conditions and handle difficult missions." He reflected further on its impact on him personally, "My training was thorough. It prepared me well for my responsibilities and duties as a combat paratrooper."[267]

10

Fort Bragg to Camp Shanks and then...England

(Once we arrived in New York), we knew for certain we were Europe-bound.
— Pat Christenson

On the 21st of July, Easy Company, and the rest of Second Battalion, formed up early at the train station outside of Camp Breckinridge for the trip back to Fort Bragg. The ride eventually covered 1,200 miles in distance and took 13 hours, spread out over the course of three days. The ride was no different from the previous ones they had taken. It was something to endure, not enjoy. Bag lunches, much like the ones they were given during the Tennessee Maneuvers, were the norm, and other than a few sinks in the lavatory cars, the train lacked facilities for the men to wash up. Once again, they were a bit ripe by the time they climbed off their train cars on the 24th of July. Trucked back to the base, the company was ushered towards barracks, this time on the main side of Fort Bragg. They wouldn't have long to settle in because training would begin in earnest in early August.

Scenes from Fort Bragg.
Left: Bill Guanere, wearing his khaki service uniform and overseas cap, points an accusatory finger at one of his platoon mates. (Author's Collection)
Middle: Burr Smith and Skip Muck pose on the steps of one of Easy's barracks buildings. Interestingly, Muck wears standard infantry leggings. (Photo courtesy of Robert Burr Smith via Susan Smith Finn)
Right: Robert T. "RT" Smith and Myron "Mike" Ranney outside the First Platoon barracks. (Author's Collection)

While Easy Company got ready for more field problems, things were cooking at the Regimental Headquarters. "Have Lieutenant Winters report to me immediately!", Colonel Sink called out to his orderly. "Yes, Sir!", the orderly replied. Grabbing the telephone handle and dialing the appropriate number, he called over to the E Company Headquarters office.

"Have Lieutenant Winters report to Colonel Sink immediately." The word was passed to Winters who dutifully reported to the Regimental HQ, wondering what the heck was so important that the Colonel needed him specifically.

"Dick, I don't like the idea of the officers looking like ragamuffins. Some of them are still wearing the rain coats they received at Toccoa. The weather's probably gonna be rough in England. They tend to get a lot of rain. I've assessed each man for the cost of a new trench-style overcoat. Also, The Officers' Mess is running low on booze and I want our supply of whiskey restocked immediately. The orderly has a list of how many coats and what sizes, plus all the money for them and the booze. He'll also have your orders authorizing you to be off base. See to it that it gets done." Winters had hardly got a word in edgewise, but he saluted smartly and exited the office.

Gathering up everything that was needed from Sink's orderly, Winters went back to the company office and informed Captain Sobel. Then it was back to the officer's quarters to gather his things. Winters then headed to the post's main gate and flagged a taxi to the train station. He had decided to head to Philadelphia, figuring he could kill two birds with one stone. A big city like Philly would definitely have the overcoats and booze, right?

Winters was only half right. He had no trouble buying the overcoats, but the whiskey was another thing entirely. Booze had been rationed and while he could have purchased a bottle or two here and there, it wouldn't be the amount that the Colonel had envisioned. Frustrated, Winters took a cab back to his home in Lancaster. Stonewalled by the lack of available booze and at the end of his rope with the lack of ideas, he called back to the Regimental HQ for guidance. "Winters, this is Lieutenant Colonel Chase." Chase, the Regimental XO, told him to stay where he was by the phone and he would get back to him. The next day, Chase was as good as his word. He gave Winters an address in New York City and the name of a man that could help with the problem. Winters would be heading to the company headquarters of Schenley Distillers Company.

The next day he hopped a train to New York and made his way to Schenley's. Ushered into a corporate office by a very attractive secretary, Winters was waved over by a corpulent older man who sat with one leg elevated on a small stool. "I understand you have a bit of problem, soldier?" the man queried. "Yes, Sir," Winters replied. "I am in need quite a bit of whiskey for our Officer's Mess. Do you think you could help me?" Winters handed over the list he had been provided by the Regimental Orderly. The man looked over the list and without batting an eye responded, "Yes, I can take care of that order." And with that affirmation, one of the other secretaries took down details while Winters sat down to endorse the pile of $20 and $50 money orders he had foolishly converted the unit funds into.[268] With the issue of the overcoats and whiskey finally put to bed, Winters found his way to Penn Station, purchased a ticket, and boarded a train for the trip back to Fort Bragg.[269]

* * *

Near the end of the month, Easy Company experienced another influx of new troopers. Many of whom would go on to become staples in the unit. On the 29th of July, John Mayer, Earl "Mac" McClung, Thomas "Mac" McCreary, Bill McGonigal, John "Jack" McGrath, Walter McKay, John McKinstry, James "Jimmy" McMahon, Arthur "Art" Mauzerall, Francis "Frankie" Mellett, Joachim "Jackie" Melo, Vernon "Menzey" Menze, Gussie Monroe, Donald "Moonbeam" Moone, Alton More, David "Baby" Morris, and Sergio Moya, all signed into Company HQ. After checking in with First Sergeant Evans, each was assigned to their new platoon.[270]

At Fort Bragg, Easy Company received several transfers.
Left: Earl McClung joined Third Platoon and became a mainstay, and eventually one of the best combat soldiers in the company. (Author's Collection)
Middle: Jack McGrath was a bit older than most of the men of Easy Company. He was quiet and dependable, two qualities that held him in high regard with the men. (Photo courtesy of the Charles Rhinehart Collection via Jay Hopkins)
Right: Don Moone (seen on the right with Alton More) joined the paratroopers because of the sharp uniforms. (Photo courtesy of Ed Mauser)

McClung would eventually develop a well-founded reputation as the best field soldier in the company. McCreary, McMahon, Melo, Moone, More and Morris would serve with Easy until the end of the war, with More becoming famous for acquiring a pair of photo albums from the Eagle's Nest in Berchtesgaden. McGonigal, Mellett, Menze, and Moya would all be killed in combat, McGonigal and Moya while aboard the ill-fated Stick #66 on D-Day. Menze would be killed during the fighting in Nuenen, Holland, and Mellett during the attack on Foy, Belgium. Several of the others would be wounded and knocked out of the war. But that was in the future. For now, much like the previous rounds of new troopers being assigned the company, these men would have to integrate, and as "Baby" Morris recalled, "get their ears blistered by the 'nuts' who were there (at Toccoa)," being constantly reminded that they weren't "Toccoa Men."[271]

At twenty-one, Jack McGrath was a bit older than many of his fellow troopers. He had originally started out as a communicator but was told that he would, in addition to his communication duties, on occasion man a bazooka in Easy's First Platoon. Jack's account of how he became a bazooka man is quite comical. "They needed a bazooka man...'What is this?' 'This is a bazooka. You do this, you do that, blahblah and that's it.' 'Okay, that sounds easy.' Now, that bazooka was the old type; it was one piece. The question I had was: How the hell do you get out the door of the plane with that? I soon learned. You bend it one way, as soon as you get out you let it go the other way. Otherwise, you break your legs. So, that's how I became a bazooka man." He also learned early on that they were required to double time everywhere they went and worked with the other 'rocket' men, like Ed Tipper, to figure out the best, and safest, way to jump the four and a half foot long, and almost 13-pound bazooka tube. He felt good about being with such a tight unit, and that they were, "full of pep, vim and vigor."[272]

Some of the men were quickly accepted. Menze was befriended by fellow Second Platoon man Rod Bain. The pair soon became very close. But it wasn't so smooth for all the new joins. While "Shifty" Powers took an instant liking to "Mac" McClung, a fellow Native-American who hailed from Inchelium, Washington, he took an instant dislike to Frankie Mellett. Powers, a southerner from the mountainous area in western Virginia, just couldn't find any common ground with Mellett, a big city "Yankee" from the Bronx, New York. McClung, along with the feuding pair, would serve together in Third Platoon.[273] Moya, from California, was immediately taken under the wing of "RT" Smith when he was assigned to the Mortar Squad in First Platoon. Smith quickly became fond of the earnest young Californian, sensing the soldierly potential in him. He would not be disappointed.[274]

* * *

The week or so after arriving at Bragg, the company entered their next training cycle, but before they kicked off with the exercises, there were some administrative tasks to be completed. First off, the non-commissioned officers were assigned to four-man rooms. This was in stark contrast to their previous living accommodations. This was followed up by sore jaws and sore arms as the men all made visits to the dentist and medical clinic for some dental work and additional immunizations. The 'molar mechanics' and 'pill rollers' had to ensure that each man was medically fit to deploy overseas. Despite the drilling and filling and the feeling of being used as a pin cushion, the men were impressed in the upgrade in food quality. Most had become accustomed to the meager servings and the bland menus that they had experienced up to this point. Instead of SOS for every breakfast, they started to see hash browns and bacon for breakfast more frequently. While still having to put up with dehydrated eggs, the other menu items did make them more tolerable.

They were also issued all new uniforms and equipment. Up to this point, many had been going on field problems wearing the same items they had received at Toccoa as part of their basic issue. Well-worn pistol belts, field suspenders, canteen carriers, entrenching tool carriers, cartridge belts and musette bags were handed over to Sergeant Murray Roberts and Private Max Clark in the Company S-4 (Supply) Section. It was a one-for-one swap at the supply 'hut,' with frayed and ragged gear being turned in, and factory-fresh equipment being handed back over the counter to each trooper in the company. Jack McGrath remembered turning in his old gear, which consisted of a lot of pre-war or WWI-vintage equipment and being issued almost all new gear and uniforms. The men also received new M42 jump suits, and new jump boots. It was like early Christmas for the paratroopers.[275]

The men also received a steady diet of weekend passes. While Sobel never wavered in his demand for adherence to standards, the men of Easy had adapted to them well. With passes in hand, the paratroopers made their way into Fayetteville, where the Town Pump Tavern became a favorite hangout. Unfortunately, they weren't the only troops in the area that felt the same way. Whether they were tankers, infantrymen or other paratroopers, on any given night, it didn't take long for one unit to be insulted by another, and fueled by beer and whiskey, the fists would start flying. Many of the Easy troopers took to sewing a silver dollar coin behind the parachute patch on their overseas caps. With that bit of metal in there, they could effectively use their caps as a blackjack in a fistfight. And use them they did.[276]

But the best way to keep the men out of trouble was to get them back to training. The communications men to include, Carl Riggs, Rod Bain, Jack McGrath, George Luz, "Gordy" Carson, "Moe" Alley, Frank Perconte, Jerry Wentzel, Bill McGonigal, Joe Liebgott, and Bill Maynard. For two weeks they learned about the intricacies of the SCR-300 man pack radio and its smaller cousin, the SCR-536 "Handy Talkie." It was an enjoyable course as it was taught "gentleman" style with no tests, and even better for the men, no Captain Sobel.

For the rest of the company, they went back to the firing ranges. Entirely comfortable using just about every weapon in the inventory, almost every man in Easy qualified as an Expert. Submachine guns like the Thompson and Grease Gun were especially enjoyable to fire. Less enjoyable was the cleaning of those weapons after firing. And even further down the list of enjoyable aspects of the weapons, was the carrying of them for long periods of time. Unfortunately, that's exactly what they would have to do. In addition to range time, the men spent time in the classrooms, learning about different subjects. Noncom Leo Boyle, after delivering several zealous presentations on chemical warfare, earned the comical nickname of, "Fearless Phosgene."

In the early hours of August 10th, the entire company stepped off on a 12-mile conditioning hike with every man carrying their assigned weapon and equipment. Five and a half hours later, the company returned, a little worse for wear. But the long marches at Toccoa had hardened Easy to a fine point. Not a single man fell out. The next day, they took part in a Regimental Review in front of Colonel Sink and his entire staff. The men looked good and marched sharply, and as a result, much of the company were given furloughs and admonished to not be late in returning. Unfortunately for him, David Morris did arrive late, and he was designated 'absent without leave' on the Company Morning Report. Upon his eventual return, he was confined to the guardhouse at Bragg. That confinement would last all the way until Easy's arrival in the ETO, or Eastern Theater of Operations. But no one at company knew that yet. All Morris knew was that he was in the proverbial doghouse.[277]

* * *

Once the men boarded, they placed their barracks bags on small overhead shelves and picked seats next to their buddies. The easiest way to kill time on a long train ride was to share it with a friend. It took several hours to board the entire Regiment, but before long, they were seated and decks of cards and pairs of dice made appearances. The 506th was headed to Camp Shanks, New York which would be their point of embarkation. As the train rumbled down the tracks, Lew Nixon sought out his friend Dick Winters. He found him sitting with Harry Welsh. "Gents, in case you haven't figured it out, we're headed to England," Nixon said with a smile.[278]

Top: The barracks buildings at Camp Shanks were much like the ones Easy had stayed in at other bases. (Photo courtesy of the Times Union of Albany)
Bottom left: A line of soldiers snakes their way towards a dockside embarkation point at Camp Shanks. (Photo courtesy of the Times Union of Albany)
Bottom right: The Camp Shanks telephone center. Many troopers made one last phone call home from this building. (Photo courtesy of the 506th Infantry Association Website)

After arriving Camp Shanks, the men exited the train and stood in company formation. Roll was called and once everyone was accounted for, they were marched to their barracks. It was a long walk and they were "loaded to the gills with equipment" according to Pat Christenson. Each of the men expected to receive a pass for a day out on the town, but they received the opposite treatment...confined to the base. A few men managed to make unauthorized trips off base but were quickly rounded up. One of the first things they did after being confined to the base was to queue up for shots. The men were given so many shots at one point, "our arms hung from our bodies like limp rope."[279]

One day, one of the noncoms came into the barracks yelling for them to fall out wearing only a raincoat and their jump boots. They were to be subjected to another physical. On the way to the medical facility, they had to march through the Women's Army Corps' barracks area. In a bit of role reversal, razzing and catcalls rained down on the Easy men from the ladies of the WAC. Afterwards, Pat Christenson and several other First Platoon men wandered over to the bulletin board outside the Company Headquarters. The unit schedule simply stated, "usual camp duties," which typically meant there would not be conducting any formal training. So, the men wandered aimlessly around camp, often going to the PX where they bought "enough razor blades, soap, towels and what have you, to last till the end of the war."[280] This was a far cry from what they were used to under the tyrannical hands of Captain Sobel and First Sergeant Evans, who normally controlled every waking minute of each day.

While at Shanks, Woodrow Robbins (left), a tall First Platoon machine gunner, and wiry Robert Marsh squared off in the boxing ring, while George Luz (right) paired up with Pat Christenson for some wrestling. (Robbins photo courtesy of the Carson Family; Marsh photo courtesy of the author's collection; Luz photo courtesy of the Luz Family)

One night, a bunch of the First Platoon boys went to the gym to blow off steam. Woodrow Robbins and Robert Marsh went at it in the boxing ring while Pat Christenson and George Luz squared off for some unarmed grappling and wrestling. Although he was 20 pounds lighter than Christenson, Luz acquitted himself well, oftentimes running circles around the bigger man. According to Pat, they came out about even, and George once again proved that "he was one of the best."[281] He would do so time and time again, especially when it mattered most.

* * *

On the 1st of September, the men of Easy were finally given what they had been waiting for...three-day passes. But it came with a caveat. They weren't allowed to wear their jump boots or any insignia that indicated they were paratroopers. If German spies were about, no one wanted to let them know the paratroopers were getting ready to head overseas. Still, the men grumbled and bitched while cutting off their patches and digging their long-forgotten Army Service Shoes out of the bottoms of their duffle bags. After shining the dull leather of the shoes and polishing the brass on their uniforms, the men were ready to hit the town.

When the men began training back at Toccoa, Captain Sobel had instituted a 'slush fund' where each man in the company would contribute a small amount of money to be used for purchasing items for the benefit of the entire unit. Knowing that radios were an inexpensive way for the men to be entertained and informed of goings-on in the world, Sobel had decided to purchase radios for the company using the slush fund. Calling Lipton into his office, Sobel explained that he had made a deal with a company that built car radios, to purchase several for Easy and Lipton was to pick them up. Lipton dutifully went to the company's shop and purchased the radios as instructed. As he was leaving, the salesman said out loud, "How are you guys going to power the radios? I hope you have some batteries." Pausing at the door to ponder the question, Lipton shrugged his shoulders and thought, "That's a problem for the captain to sort out."[282]

The men spent the bulk of their three-day passes out in New York City. Most had never visited the city, and enamored with the big buildings and bustling nightlife, used months of accumulated jump pay to really enjoy themselves. Many rented hotel rooms rather than head back to Camp Shanks. It seemed that sleeping between clean sheets on a hotel bed was preferred over doing so in a crowded, noisy, Army barracks. On the last day of his pass, Pat Christenson really overdid it with whiskey and beer. He had managed to wobble back to Shanks, and to the barracks, finally making his way to the toilet. The room began to spin as Pat emptied his guts. With vomit staining his dress shirt, he mercifully passed out. "Bull" Randleman, also back from pass, noticed Pat laying in a heap near the toilet. Feeling bad for his friend, "Bull" picked Pat and carried him over to his bed. Once he removed Pat's shoes and tie, "Bull" tucked Pat in with almost childlike gentleness. Pat woke up the next day with a massive hangover. As he rubbed his aching head, one of the other men told him what "Bull" had done. Randleman could have easily lorded over Pat and embarrassed him over what had happened, but that wasn't in "Bull's" nature. In fact, "Bull" never mentioned a word of the incident to Pat. As a result, Pat vowed to never forget his friend's kindness.[283]

"Dull" Randleman (left) looked after Pat Christenson (middle) after the latter had a few too many one night on pass. "Moose" Heyliger (right) found out about the birth of his son as he was boarding the RMS Samaria. (Randleman photo courtesy of the author's collection; Christenson photo courtesy of the Christenson Family; Heyliger photo courtesy of the author's collection)

The men were awakened early on the 5th of September for their manifest call. Hurry up and wait was the order of the day. Finally, they marched single file up the gangplank. Names were checked off on a clipboard before boarding. They were to be transported to England by the *RMS Samaria.* They wore their wool uniform, with overseas cap, and jump boots. Around their waists, they wore their pistol belts with canteen and first aid kit attached. They were also issued an inflatable lifebelt in the event the boat was torpedoed and sank, or if they fell overboard. Despite that precaution, few of the men believed they'd survive in the Atlantic for very long should either event occur. Outside of that, they carried most of their worldly possession in their "A" barracks bag. Their duffle bags with the remainder of their equipment and clothing had been palletized and loaded into the ship's hull.[284] [285] Christenson said they looked and felt like, "a colony of ants", but the process was orderly, lacking the confusion of similar type movements they had experienced in the past.[286]

While the reality of the moment was not lost on the men as they boarded, one Easy Company officer was in good spirits. As he was walking up the gangplank, Third Platoon Leader, Frederick "Moose" Heyliger, was informed that his wife had given birth to a son. Heyliger whooped it up, hollering all the way onto the ship, and the other officers congratulated him with handshakes and slaps on the back in honor of the birth of "Little Moose." As he boarded the Samaria, Moose began signing songs, and his singing lasted well into the evening.[287]

The S.S. Samaria, later requisitioned by the Royal Navy and rechristened the RMS Samaria, carried Easy Company and the rest of the 506th Parachute Infantry Regiment across the Atlantic to England. (Photo courtesy of the website, Norway Heritage)

The *Samaria* was a 600-foot Cunard Line transatlantic ocean liner. She had been launched in November 1920 and entered service in April 1922. The Samaria had been an S.S. ship that plied the Liverpool to Boston and New York routes. She was requisitioned by the Royal Navy in 1941 and rechristened the RMS Samaria. From that point on she performed troopship duties. With two screws, the *Samaria* could sail at 16 knots. Normally, she could carry approximately 2500 passengers and crew, but since being pressed into service, her carrying capacity had been increased out of necessity.[288] She would now carry the entire 506th Parachute Infantry Regiment to England, a total of 2363 paratroopers, along with 434 crew members to England.

The enlisted men were assigned spaces below decks, while the officers shared state rooms. Space was at a premium. The enlisted were assigned two or three to a Navy style rack, with the racks stacked three high. Outside of hanging the barracks bag from the racks, each man possessed little to no personal space. For the officers, it was four to a state room with a tiny common area in the center of the room. The four men had to share a single desk or table on which they could write letters or play cards. The below spaces were poorly ventilated and many of the men would spend as much time up on deck as possible over the coming ten-day trip.

Almost uniformly, the men of Easy Company made their way up to the top deck. As the *Samaria* slowly pushed away from the pier and out into New York Harbor, the men lined up along the rails to look at the Statue of Liberty. Most were lost in thought, wondering if they'd live to see this same view on the return voyage. Only time would tell.

Space was limited on the Samaria. For the voyage carrying the 506th, she carried almost 400 more people than was intended. Left: The top deck was standing room only during the voyage. (Photo courtesy of the 506th Infantry Association Website) Right: One of the Samaria's lifeboats is clearly visible behind Audrey Lewallen (left) of the Light Machine Gun Platoon, 3rd Battalion HQ, 506th Parachute Infantry. (Photo courtesy of the 506th Infantry Association Website)

Appendix I: The First to Arrive: The Toccoa Men

A note as you peruse the roster list of E Company troopers. The use of FNU means 'first name unknown'. Despite rigorous research, information on some of these troopers was severely lacking. Without Army Service Numbers (ASN) and sometimes lacking first names due to the inadequacy of early handwritten morning reports, it was very difficult to locate details on those troopers.

Outside of that, I added relevant information that wasn't necessarily going to be covered in the narrative, which includes location of first noted duty station for each trooper, and the range of ranks that soldier had attained during that time frame. If the trooper does not have a platoon assignment following his name, it's because that information is unknown at the time of this writing.

20 July 1942 – 1 December 1942

Company Headquarters

2/LT-1/LT WALTER "WALLY" MOORE (ASN: O-454603) - Commanding Officer/Platoon Leader - Moore was the first CO, holding the position from 03AUG42 - 07AUG42. After being relieved as CO, Moore stayed on as the 3rd Platoon Leader. On 23AUG42, Moore was sent to the Parachute School on detached service orders.

2/LT GROVER L. SMITH (ASN: O-403343) - Commanding Officer/Platoon Leader - Relieved Moore as CO. Smith's tenure may have been one of the shortest in the history of the US Army, literally 2 days. Even after being relieved, Smith remained with the company as a platoon leader.

2/LT JAMES L. DEAN (ASN: O-385071) - Commanding Officer/Platoon Leader - Dean was initially brought to E Co as a platoon leader on 07AUG42 but ended up relieving Smith as CO on 09AUG42. He lasted two days as well, being relieved as CO by Sobel. He would remain with E Co until 5AUG42, when he was transferred to 2HQ/506.

1/LT/CPT HERBERT M. "THE BLACK SWAN" SOBEL (ASN: O-304109) - Commanding Officer - Sobel's arrival on 11AUG42, signaled the end of the revolving door of CO's for E Co. He was promoted to the rank of captain on 29NOV42.

1/LT ROY RILEY (ASN: O-450508) - Company Executive Officer - Assigned to E Co on 15AUG42 and was transferred out of the 506th on 14SEP42.

1/LT CLARENCE HESTER (ASN: O-425255) – Company Executive Officer - Hester was assigned to E Co as the Executive Officer on 15SEP42.

1/SGT ALBERT H. "AL" MILLER JR. (ASN: 6372713) - Assigned to E Co on 20JUL42 and was reassigned to RHQ/506 on 27JUL42.

1/SGT FNU MORRIS (ASN: UNK) - Assigned to E Co on 27JUL42.

S/SGT FNU BIRMINGHAM (ASN: UNK) - Assigned to E Co on 20JUL42.

S/SGT FNU LANE (ASN: UNK) - Assigned to E Co on 20JUL42.

S/SGT FNU LONG (ASN: UNK) - Assigned to E Co on 20JUL42.

Company Training Cadre

S/SGT - 1/SGT HARVEY H. MOREHEAD (ASN: 6396532) - Company Cadre 1SG– Company Cadre 1SG - Photographic evidence indicates that Morehead made the march from Toccoa to Ft. Benning. William Evans likely took over as first sergeant after the company arrived at the Parachute School. Morehead remained with E Company at least until Feb. 1943 as his name appears in a morning report from that period. To remain with the 506th, Morehead eventually took a reduction in rank to S/SGT and was reassigned to F Company.

SGT – S/SGT JOE E. STEDMAN (ASN: 6859578) – 1st Platoon Cadre SGT - At one point, Stedman was the middleweight boxing champion of the 101st Airborne.

PFC – S/SGT NORMAN A. FORDE (ASN: 7032054) – 2nd Platoon Cadre SGT – 2nd Platoon Cadre SGT - Forde was busted down to PFC for going AWOL at one point. He gradually made his way back up and was promoted to S/SGT on 21NOV42.

SGT – S/SGT STEVEN KUDLA (ASN: 13006511) – 3rd Platoon Cadre SGT

Officers

2/LT WARREN W. CANNRIGHT (ASN: O-430260) – Platoon Leader - Cannright joined E Co on 08AUG42 and was transferred out of the 506th on 14SEP42.

2/LT VINCENT S. TABOR (ASN: O-382540) – Platoon Leader - Tabor was assigned to E Co on 20AUG42 and transferred out on 03NOV42.

2/LT FREDERICK "MOOSE" HEYLIGER (ASN: O-1010235) – 3rd Platoon, Assistant Platoon Leader - Heyliger was assigned to E Co on 20AUG42 and was quickly transferred to 81mm Mortar Platoon/2HQ/506 on 25AUG42. He would return to the Company in April 1943.

2/LT-1/LT RICHARD D. "DICK" WINTERS (ASN: O-1286582) – 2nd Platoon Leader - Winters joined E Co on 25AUG42. He later took over First Platoon. Promoted to First Lieutenant on 26OCT42. After college, Winters first served as an enlisted man while stationed at Camp Croft and reached the rank of corporal prior to going to OCS at Fort Benning in early 1942. After graduation, he was assigned to Camp Croft again as a Second Lieutenant while awaiting his permanent duty assignment which eventually was Camp Toombs (aka Camp Toccoa) GA. Winters was CO of Second Platoon while at Toccoa and had First Platoon as well prior to making First Lieutenant when he became E/506 XO.

2/LT-1/LT SALVE H. "MAT" MATHESON (ASN: O-444758) - 1st Platoon, Assistant Platoon Leader - Assigned to E Co on 28AUG42. Promoted to 1/LT on 29NOV42 and transferred to 2HQ/506. According to Pat Christenson, Matheson was placed in charge of the machine gunners while at Toccoa.

2/LT LEWIS "BLACKBEARD" NIXON (ASN: O-1286540) - 2nd Platoon, Assistant Platoon Leader - Nixon was assigned to E Co on 19SEP42.

2/LT EDGAR W. HOUSEMAN (ASN: O-1283609) - Platoon Leader/Assistant Platoon Leader - Houseman was assigned to E Co on 24SEP42, as a platoon leader or assistant platoon leader, and was transferred out of the company on 03NOV42.

2/LT LEROY C. SANO (ASN: O-415664) - Platoon Leader/Assistant Platoon Leader - Sano was assigned to E Co on 03OCT42 and by May '43 had been assigned to 2HQ/506. Little is known about his time in the 506th but he later went on to serve with distinction in B & C, 551st Parachute Infantry.

2/LT-1/LT JAMES K. DAVIS (ASN: O-404800) – Platoon Leader/Assistant Platoon Leader - Davis was assigned to E Co on 06OCT42 and was reassigned to 2HQ/506 on 09OCT42. He came back to E Co in August '43. He was featured in one of the Atlanta-Constitution Journal photos from the march to Atlanta where he was depicted along side Dick Winters, directing the employment of a machine gun section from First Platoon.

2/LT GEORGE LAVENSON (ASN: O-1286510) – 3rd Platoon, Assistant Platoon Leader - Lavenson was assigned to E Co on 09OCT42.

2/LT GEORGE A. KAHDY (ASN: O-1206052) – Platoon Leader/Assistant Platoon Leader - Kahdy was assigned to E Co on 19OCT42.

2/LT FREDERICK E. HUTTO (ASN: O- 341250) - Hutto was assigned to E Co on 21OCT42. Not much is known about his time in the 506th but apparently Hutto was captured by the Germans and spent time in Stalag 9B Bad Orb Hessen-Nassau, Prussia.

2/LT ARCHIBALD S. BARNWELL (ASN: O-447152) – 1st Platoon, Assistant Platoon Leader - Barnwell was assigned to E Co on 06NOV42.

2/LT WARREN R. "FARTSACK" ROUSH (ASN: O-1290664) – 1st Platoon & 2nd Platoon, Platoon Leader/Assistant Platoon Leader - Roush was assigned to E Co on 06NOV42 as an assistant platoon leader with First Platoon. He eventually ended up commanding Second Platoon.

The Enlisted Men (in order of arrival)

25 AUGUST 1942

PVT WILFRED H. BROWN (ASN: 13066350) - Transferred to 3HQ/506 on 22SEP42.

PVT/CPL JOSEPH STICKLEY (ASN: 6944370)

PVT JAMES F. "STINKY" COLEMAN JR. (ASN: 13066282) – 1st Platoon

PVT/CPL HARRY S. STANFIELD (ASN: 13066267) - Stanfield was busted from CPL to PVT and transferred to the 78th Infantry Division.

PVT/PFC DARRELL C. "SHIFTY" POWERS (ASN: 13066266) – 3rd Platoon, 2nd Squad

PVT ROBERT E. "POPEYE" WYNN (ASN: 13066265) – 3rd Platoon, 1st Squad

PVT HAROLD A. SOUTHWORTH (ASN: 17078657) - Transferred to 3HQ/506 on 27AUG42.

PVT/PFC CLARENCE M. "CRASH" TRIDLE (ASN: ASN: 17058836) – 3rd Platoon Mortars

PVT FORREST L. "GOODIE' GUTH (ASN: 13099211) – 3rd Platoon, 1st Squad

PVT/T-5 LEDLIE RIDLEY "PARK" PACE (ASN: 13099192) – 3rd Platoon, 1st Squad MG (AG)

PVT CHESTER R. ESCHENBACK (ASN: 13099212)

PVT LYNN H. LINES (ASN: 13091306)

PVT ALLEN R. PHILLIPS JR. (ASN: 13100793) - Transferred to 2HQ/506 on 15SEP42.

PVT/PFC KENNETH T. BALDWIN (ASN: 13099345) – 3rd Platoon, 1st Squad

PVT/CPL/SGT AMOS J. "BUCK" TAYLOR (ASN: 13099363) – 3rd Platoon, 2nd Squad

PVT CARL L. "DUTCH" FENSTERMAKER (ASN: 13099212) – 3rd Platoon, 1st Squad

PVT/PFC RODERICK G. "ROD" STROHL (ASN: 13099213) – 3rd Platoon Mortars

PVT NEILON B. GLASS (ASN: 13091574)

PVT JOSEPH P. WHITECAVAGE (ASN: 13100781) – Benning, Battalion Mess - Whitecavage appeared on the 1942 2nd Battalion Christmas Roster. He later transferred to 2nd Bn HQ as a Battalion Cook.

PVT/PFC WALTER S. "SMOKEY" GORDON (ASN: 13099280) – 3rd Platoon, 1st Squad MG

PVT FNU LANDHERR (ASN: UNK) - Transferred to 3HQ/506 on 27AUG42.

PVT WILLIAM "WILLIE" WAGNER (ASN: 13099382) – 3rd Platoon

PVT STANLEY W. WOJCIECHOSKI (ASN: 32482214) - Transferred to 3HQ/506 on 26AUG42.

PVT HERMAN "HANK" HANSON (ASN: 16100573) – 2nd Platoon, 1st Squad - Herman Hanson's other nickname was "Herman the German."

PVT WILLIAM M. PATTERSON (ASN: 16100611) – 1st Platoon, Mortar Squad, Ammunition Bearer - Transferred to 2HQ/506 on 18NOV42. He would return to the company after Bastogne.

PVT FRANK J. "PERCO" PERCONTE (ASN: 16100572) – 1st Platoon Comm

PVT DANIEL B. "DANIEL BOONE" WEST (ASN: 15116953) – 3rd Platoon, 1st Squad MG

PVT/CPL CLIFFORD CARWOOD "LIP" LIPTON (ASN: 15116961) – 3rd Platoon, 1st Squad

PVT WAYNE A. "SKINNY" SISK (ASN: 15116959) – 3rd Platoon, 1st Squad

PVT GORDON W. ALLEE (ASN: 17073586) - Transferred to 3HQ/506 on 14OCT42.

PVT PAUL C. "HAYSEED" ROGERS (ASN: 17058810) – 3rd Platoon Mortars

PVT JOHN G. CONNELL (ASN: 13111956) – 3rd Platoon, 1st Squad

PVT ALBERT MAHMOOD (ASN: 17107651)

PVT GORDON L. NEUENFELDT (ASN: 17107823)

PVT/CPL MYRON "MIKE" RANNEY (ASN: 17107816) – 1st Platoon, HQ & 2nd Squad

PVT DONALD F. BARLOW (ASN: 16073437) - Transferred to 3HQ/506 on 28SEP42.

PVT FNU LUDWIG (ASN: UNK) - Transferred to 3HQ/506 on 27AUG42.

PVT FNU STEVENS (ASN: UNK) - Transferred to 3HQ/506 on 27AUG42.

PVT EUGENE W. HEISER (ASN: 12095633) - Transferred to 3HQ/506 on 09SEP42.

29 AUGUST 1942

PVT CHARLES "CHUCK" GRANT (ASN: 19126213) – 2nd Platoon, 2nd Squad

PVT BENJAMIN J. "BUD" STONEY (ASN: 39530033) – 1st Platoon

PVT FNU LEWIS (ASN: UNK) - Transferred to 3HQ/506 on 31AUG42.

PVT/CPL GEORGE W. THOMASON (ASN: 39530136) - Thomason was busted from CPL to PFC on 16NOV42.

PVT EARL W. HINDS (ASN: 1912683) - Transferred to 3HQ/506 on 03SEP42.

PVT SIDNEY A. LUSH (ASN: 39531803) - Transferred to 3HQ/506 on 03SEP42.

PVT/CPL/SGT ELMER L. "MOE" MURRAY JR (ASN: 19126297) – Company HQ

PVT CAMPBELL T. "CT" SMITH (ASN: 39531483) – 3rd Platoon Mortars

PVT/CPL/SGT TERRENCE "SALTY" HARRIS (ASN: 19091437) – 3rd Platoon

PVT DEAN R. SCOTT (ASN: 39848537) - Transferred to 3HQ/506 on 03SEP42.

PVT RICHARD M. "RED" WRIGHT (ASN: UNK) – 3rd Platoon, 1st Squad

PVT EDWARD L. KEIZER (ASN: 19126291) - Transferred to 3HQ/506 on 03SEP42.

PVT WILLIAM S. "CHICO' METZLER (ASN: 19126206) – 1st Platoon, 2nd Squad

PVT ARTHUR "ART" YOUMAN (ASN: 39251325) – 3rd Platoon, 2nd Squad

31 AUGUST 1942

PVT/CPL ROBERT B. "BURR" SMITH (ASN: 12131573) – 1st Platoon, 1st Squad/Company HQ - Burr Smith was busted from CPL to PVT on 16NOV42.

PVT/CPL/SGT TASKEL ELLIS (ASN: 12082030) – 1st Platoon

PVT/PFC GORDON "GORDY" CARSON (ASN: 12130792) – 1st Platoon Comm

PVT RICHARD F. BERG (ASN: 12131622) – 1st Platoon, Benning - Berg appeared on the 1942 2nd Battalion Christmas Roster. He was transferred in Feb 1943 to Casual Company, Parachute School, Ft. Benning, GA.

PVT RICHARD L. ELLWANGER (ASN: 12130827) - Transferred to 3HQ/506 on 16SEP42.

PVT JAMES SHEREMETA (ASN: 12130828) - Transferred to the 78th Infantry Division on 17NOV42.

PVT/PFC WARREN H. "SKIP" MUCK (ASN: 12131169) – 2nd Platoon Mortars

PVT EARL W. WAGNER (ASN: 13107569) - Transferred to 3HQ/506 on 01SEP42.

PVT JOHN H. ADAMS (ASN: 14137822) - Transferred to 3HQ/506 on 01SEP42.

PVT ROBERT E. JARRETT (ASN: 14137845)

PVT JAY S. DICKERSON (ASN: 14137803) – 2nd Platoon Mortars

PVT/PFC DEWITT "ALABAMA" LOWREY (ASN: 14131458) – 2nd Platoon MG

PVT OWEN L. ANDREWS (ASN: 15107705)

PVT/CPL PAUL L. BECKER (ASN: 15107710) – Toccoa - Becker appeared on the 1942 2nd Battalion Christmas Roster.

PVT/PFC/CPL JAMES V. "RED" BENTON (ASN: 15107638) – 1st Platoon, 2nd Squad

PVT JAMES "JIM" CAMPBELL (ASN: 15107892) – 2nd Platoon, 1st Squad

PVT JOSEPH M. JORDAN (ASN: 15107761) – 1st Platoon Mortars

PVT/PFC JAMES W. MARTIN (ASN: 15107902)

PVT ROBERT L. MAXWELL (ASN: 15107683)

PVT JOHN L. SHEEHY (ASN: 12131335) – 2nd Platoon, 2nd Squad

PVT JOSEPH W. SYWULSKI (ASN: 12130829) - Transferred to 3HQ/506 on 14OCT42.

PVT WILLIAM D. "BILL" SERILLA (ASN: 16109180) – 2nd Platoon

PVT DENVER "BULL" RANDLEMAN (ASN: 16085544) – 1st Platoon, 1st Squad

PVT/CPL/SGT JAMES L. "PUNCHY" DIEL (ASN: 16100615) – 2nd Platoon

PVT BRUCE A. HUDGENS (ASN: 16100406) – 2nd Platoon – The tough looking Hudgens was originally from Chicago, IL.

PVT/PFC JOHN W. ROSSMAN (ASN: 13089375)

PVT GEORGE J. RAJNER (ASN: 15107784) – 1st Platoon

PVT LAVON P. REESE (ASN: 15107830) – 3rd Platoon

PVT GERALD D. SNIDER (ASN: 15107810) – Company HQ Cook

PVT FLOYD "TAB" TALBERT (ASN: 15107832) – 3rd Platoon, 1st Squad

PVT/PFC RODERICK G. "ROD" BAIN (ASN: 19114638) – 2nd Platoon Comm

PVT/CPL LEO D. "FEARLESS PHOSGENE" BOYLE (ASN: 39315440) – 1st Platoon

PVT DALE G. CANNON (ASN: 39313991) - Transferred to 3HQ/506 on 01SEP42.

PVT BERNARD "BARNEY" CUNNINGHAM (ASN: 39185303) – 2nd Platoon, 1st Squad

PVT GEORGE E. DURFEE (ASN: 19123033) - Transferred to 3HQ/506 on 16SEP42.

PVT THOMAS D. FLEWIN (ASN: 39185262) - Transferred to 3HQ/506 on 22SEP42.

PVT LEONARD O. LONG (ASN: 39314289) - Transferred to 3HQ/506 on 01SEP42.

PVT ROBERT F. "BOB" MANN (ASN: 39185214) – 3rd Platoon

PVT JOHN "JOHNNY" PLESHA (ASN: 39185305) – 2nd Platoon, 2nd Squad MG

PVT ORVIL S. STAIR (ASN: 39314895) - Transferred to 3HQ/506 on 01SEP42.

PVT PAUL J. SULLIVAN (ASN: 39314228)

PVT RAY BREEDLOVE (ASN: 39185249) - Transferred to 3HQ/506 on 01SEP42.

PVT THOMAS "TOMMY" BURGESS (ASN: 39185186) – 2nd Platoon

PVT ALFRED D. CLARK (ASN: 14137886) - Transferred to 3HQ/506 on 02SEP42.

PVT FRANK B. "COWBOY" GRANT (ASN: 39184975)

PVT EVAN R. DAVIES (ASN: 39185183) - Transferred to the 78th Infantry Division on 17NOV42.

PVT/CPL JOHN F. FIEGUTH (ASN: 39185220) – 3rd Platoon

PVT CHARLES D. GARFIELD (ASN: 39314023) - Transferred to 3HQ/506 on 09OCT42.

PVT VERN R. LARRANCE (ASN: 39313959) - Transferred to 3HQ/506 on 09SEP42.

PVT STANLEY I. LURVEY (ASN: 19114697) - Transferred to 3HQ/506 on 09SEP42.

PVT IRA D. MCCRACKEN (ASN: 19114702) - Transferred to 3HQ/506 on 03SEP42.

PVT JOHN P. SHEELEY (ASN: 39314111) – 1st Platoon, 1st Squad

PVT ROBERT T. STORCH (ASN: 39185167) - Transferred to 3HQ/506 on 18SEP42.

PVT/PFC ROBERT "RIP" VAN KLINKEN (ASN: 39186454) – 1st Platoon, 1st Squad

PVT/PFC CONRAD M. BAAY (ASN: 39186530) – 3rd Platoon

PVT MARSHALL C. CLARK (ASN: 38185198) - Transferred to 3HQ/506 on 01SEP42.

01 SEPTEMBER 1942

PVT PHILIP "PHIL" PERUGINI (ASN: 12127453) – 3rd Platoon, 1st Squad

PVT THEODORE R. DAVIS (ASN: 39097975) - Transferred to 3HQ/506 on 08OCT42.

PVT LAUREL W. SISTLER (ASN: 17075869) - Transferred to the 78th Infantry Division on 17NOV42.

PVT ORA M. CHILDERS (ASN: 18137423) – Benning - Childers appeared on the 1942 2nd Battalion Christmas Roster. He was assigned to Parachute Class #49 (December 14, 1942, to January 2, 1943).

PVT/PFC MURRAY B. ROBERTS (ASN: 14083048) – Company HQ

PVT WALTER E. MARTIN (ASN: 16109322)

PVT EDWARD J. "ED" TIPPER (ASN: 16106128) – 1st Platoon & 3rd Platoon, 1st Squad - By Ed's own recollection, he started with 1st Platoon in August of 1942 and a year later he transferred to 3rd Platoon.

PVT WALTER "WALLY" WENTZEL (ASN: 16109311) – 1st Platoon - Wentzel joined Easy Company of the 506th PIR in the summer of 1942 at Camp Toccoa.

PVT PEERLESS S. JEFFERSON (ASN: 18131024) - Transferred to 3HQ/506 on 14OCT42.

PVT ADOLPH M. HELTON (ASN: 18130995) - Transferred to 3HQ/506 on 14OCT42.

PVT/PFC GEORGE E. JAMES (ASN: 18160254) - James was transferred to "Station Company" at Camp Mackall on 22MAY43.

PVT/CPL DONALD B. "HOOBS" HOOBLER (ASN: 20508303) – 1st Platoon, 2nd Squad - Hoobler was busted from CPL to PVT on 16NOV42. Hoobler enlisted the Ohio National Guard, serving with the 37th Division. He trained at Camp Shelby, Mississippi and was later discharged in November 1941, one month before Pearl Harbor was attacked by Japan. By July 1942, Hoobler had reenlisted in the Army and his basic training took place at Camp Breckenridge, Kentucky, before he volunteered with childhood friends Robert Rader and William Howell for paratrooper training. (Information courtesy of: Weyrich, Carleta, Hoobler fights in WWII elite 'Easy Company', The People's Defender, West Union, OH, April 2007)

PVT WILLIAM "SHEP" HOWELL (ASN: 20508273) – 1st Platoon, 1st Squad

PVT/CPL ROBERT J. "ROOK" RADER (ASN: 20508327) – 1st Platoon, 1st Squad

PVT J. EVANS (ASN: UNK) - Transferred to 3HQ/506 on 02SEP42.

PVT/CPL/SGT WILLIAM S. "STANTON" EVANS (ASN: 18117497) – Company HQ

PVT JOHN O. GARCIA (ASN: 18155336) - Transferred to 3HQ/506 on 09OCT42.

PVT FNU PADILLA (ASN: UNK) - Transferred to 3HQ/506 on 02SEP42.

PVT FRANK B. BOWDEN (ASN: 39840254) - Transferred to 3HQ/506 on 02SEP42.

PVT RICHARD R. GARROD (ASN: 39097956)

PVT/PFC ROBERT T. "RT" SMITH (ASN: 19087366) – 1st Platoon Mortars

PVT R. H. DAVIS (ASN: UNK) - Transferred to 3HQ/506 on 02SEP42.

PVT RUDOLPH SANDOVAL (ASN: 39097919) - Transferred to 3HQ/506 on 26SEP42.

PVT HENRY E. NELSON (ASN: 39393230)

PVT/PFC PAUL WAGNER (ASN: 39393314) – Benning - Wagner appeared on the 1942 2nd Battalion Christmas Roster. He was charged with being AWOL on 2 Feb 1943.

PVT RITCHIE CRANSTON (ASN: 15326593) - Transferred to 3HQ/506 on 02SEP42.

PVT WALTER TODD (ASN: 12131770)

PVT WARREN C. HUNTLEY (ASN: 34036001) – Benning - Huntley appeared on the 1942 2nd Battalion Christmas Roster. He was charged with being AWOL on 2 Feb 1943 (reduced from PFC to PVT).

PVT FNU BIRDSONG (ASN: UNK) - Transferred to 2HQ/506 on 02SEP42.

PVT FNU THOMPSON (ASN: UNK) - Transferred to 2HQ/506 on 02SEP42.

PVT WALTER G. HOWARD JR. (ASN: 34315821)

PVT GREGORY C. ROTELLA (ASN: 34036333) – 2nd Platoon

PVT L. C. PLUMMER (ASN: 3403680) - Transferred to 2HQ/506 on 02SEP42.

PVT BILLY S. RARY (ASN: 34591998) - Transferred to 3HQ/506 on 16SEP42.

PVT ALBERT "ALBIE" BLITHE (ASN: 13111948) – 1st Platoon, 2nd Squad

PVT ROBERT J. "BOB" BLOSER (ASN: 13099636) – 1st Platoon, 1st Squad

03 SEPTEMBER 1942

PVT MAXWELL M. "MAX" CLARK (ASN: 11087989) – Company HQ

PVT PAUL J. SULLIVAN (ASN: 39314228) – Toccoa

PVT/PFC GEORGE "PORTAGEE" LUZ (ASN: 11096443) – 1st Platoon Comm

PVT GEORGE W. WATERMAN (ASN: 11096445) - Transferred to 3HQ/506 on 26OCT42.

PVT SAMUEL M. CAWTHON (ASN: 18109447)

PVT URBAN M. SHIRLEY (ASN: 18109450) – 3rd Platoon, 1st Squad - Shirley is featured in a photo in Carwood Lipton's collection showing members of 1st Squad marching in formation. Lipton annotated the names of each trooper. Shirley was a native of Tyler, TX according to his enlistment record.

PVT/PFC WILLIAM H. "DUKE" DUKEMAN (ASN: 19087501) – 1st Platoon, 2nd Squad

PVT EVERETT J. GRAY (ASN: 19087506) – 1st Platoon, 2nd Squad MG

PVT/PFC BURTON P. "PAT" CHRISTENSON (ASN: 19087456) – 1st Platoon, 1st Squad MG

PVT JACK O. GINN (ASN: 39099876) – 1st Platoon, 2nd Squad

PVT LEROI R. GARBERG (ASN: 19087502) - Transferred to Camp Toccoa base staff on 24NOV42.

PVT WILLIAM T. "BILL" WINGETT (ASN: 19087381) – 1st Platoon, 2nd Squad MG

PVT EDWARD J. DONOHUE (ASN: 19087441)

PVT RUSSELL E. HOLFORD (ASN: 19087410) - Transferred to 3HQ/506 on 02OCT42.

PVT/CPL/SGT SHERMAN M. IRISH (ASN: 39099887)

04 SEPTEMBER 1942

PVT/PFC ELMER N. SCHUYLER (ASN: 13023556) – 3rd Platoon

PVT FNU GUTIERREZ (ASN: UNK) - Transferred to 3HQ/506 on 15SEP42.

PVT FNU HERRINGTON (ASN: UNK) - Transferred to 3HQ/506 on 07SEP42.

PVT/PFC JOSEPH D. "JOE" TOYE (ASN: 13026128) – 2nd Platoon, 2nd Squad

08 SEPTEMBER 1942

PVT JOHN O. BAY (ASN: 16095339) - Transferred to the 78th Infantry Division on 17NOV42.

PVT FREDERICK G. BEALKE (ASN: 13102819) – 2nd Platoon

PVT/PFC WILLIAM J. "WILD BILL" GUARNERE (ASN: 13113070) – 2nd Platoon Mortars

29 SEPTEMBER 1942

PVT ROBERT GREER (ASN: 28643990) - Transferred to 3HQ/506 on 30SEP42.

PVT WILLIAM F. "BILL" KIEHN (ASN: 19114027) – 3rd Platoon, 2nd Squad

PVT JAMES R. MCKINSEY (ASN: 39317347) - Transferred to the 78th Infantry Division on 17NOV42.

PVT JERRY A. WENTZEL (ASN: 19112534) – 3rd Platoon & Company HQ Comm

PVT ORAN V. HOLBROOK (ASN: 39316313)

PVT DONALD G. "MALARK" MALARKEY (ASN: 39316320) – 2nd Platoon Mortars

PVT FNU TAPPE (ASN: UNK) - Transferred to 3HQ/506 on 30SEP42.

PVT WILLIAM J. COSTA (ASN: 39102354) - Transferred to 3HQ/506 on 21OCT42.

PVT VERLIN V. "SADIE" HAWKINS (ASN: 39395642) – 1st Platoon

PVT LARRY G. GILES (ASN: 19138009) – Toccoa

PVT COBURN "JOHNNY" JOHNSON (ASN: 19138010) – 1st Platoon, 1st Squad

PVT WILLIAM J. REYNOLDS JR. (ASN: 39395650) - Transferred to 3HQ/506 on 14OCT42.

PVT JAMES H. "MOE" ALLEY (ASN: 18166060) – 3rd Platoon Mortars

06 OCTOBER 1942

PVT ROY A. GIFT (ASN: 39103550) - Transferred to 3HQ/506 on 24OCT42.

PVT JOSEPH "JOE" LIEBGOTT (ASN: 39103809) – Co HQ Comm, 1st Platoon Comm & 3rd Platoon

PVT ADAM NOBRIGA (ASN: 39103821) - Transferred to 3HQ/506 on 14OCT42.

PVT FNU FIGUEIRA (ASN: UNK) - Transferred to 3HQ/506 on 11OCT42.

PVT WOODROW W. "CURLY" ROBBINS (ASN: 39103557) – 1st Platoon, 1st Squad MG

PVT LEOPOLLO P. "ALEX" CARRILLO (ASN: 39857283) – 2nd Platoon, 2nd Squad

PVT DAVID L. HALTOM (ASN: 39396433)

PVT WILLIAM C. "BILL" MAYNARD (ASN: 39397626) – 3rd Platoon Comm

PVT JOSEPH "JOE" RAMIREZ (ASN: 39103622) – 3rd Platoon, 1st Squad Rockets

PVT CARL N. RIGGS (ASN: 39103579) – Company HQ Comm

PVT JERRY G. YOUNG (ASN: 39396382)

12 OCTOBER 1942

PVT IRA E. FARLEY (ASN: 39103349) - Transferred to 3HQ/506 on 29OCT42.

PVT FNU LEMON (ASN: UNK) - Transferred to 3HQ/506 on 24OCT42.

20 OCTOBER 1942

PVT DENZIL W. SPACE (ASN: 19096045) - Transferred to Med Det/506 on 21OCT42.

27 OCTOBER 1942

PVT JACK B. SMOTHERS (ASN: 19126155) - 1st Platoon - Smothers appears in "Gordy" Carson's journal notes of 'buddies' but not in the Pulles Roster. He was possibly a cadre member. He enlisted on 15 August 1942 out of Los Angeles, CA and was immediately assigned to the infantry. He was born in 1921 in Nebraska and was single at the time of his enlistment. He was apparently briefly assigned to the company and was released from the company rolls on 19 Feb. 1943 when he was sent to Service Company, 506th Prcht Inf.

UNDATED ARRIVAL

PVT FRANK NMI SMITH (ASN: 6966657) - Toccoa

PVT JON J. GAMON (ASN: 31051594) - Toccoa

PVT MICHAEL E. RAUSEO (ASN: 11047873) - Toccoa

PVT LEONARD J. WOODRUFF (ASN: 18062507) - Toccoa

PVT BOB H. ABERNATHY (ASN: 38315937?) - Toccoa

PVT ALTON B. AGEE (ASN: 34377357) - Toccoa

PVT ROBERT NMI ANDREAS (ASN: 34249324) - Toccoa

PVT JAMES A. CARUSO (ASN: 38312??3) - Toccoa

PVT EVERETT L. KENT (ASN: 19173856) – Toccoa

PVT BENJAMIN J. LAVENDA (ASN: 13025529) – Toccoa

PVT MAURICE G. MANTOCK (ASN: 35566730) – Toccoa

PVT JOHN A. MENIGHAN (ASN: 15338109) – Toccoa

PVT ROBERT B. NEWHART (ASN: 15335094) – Toccoa

PVT DAROLD L. STRINGFELLOW (ASN: 14084310) – Toccoa

PVT VERNON Y. TAYLOR (ASN: 19171833) – Toccoa

PVT RALPH H. THOMAS (ASN: 18192006) – Toccoa

PVT OLLIE O. TOUCHSTONE JR. (ASN: 39684349) – Toccoa

PVT TYRUS K. WILLIAMSON (ASN: 18181832) – Toccoa

Appendix II: The Roster, from Fort Benning to Camp Shanks

Company Headquarters

1/LT-CPT HERBERT SOBEL (ASN: O-304109) - Company HQ, CO (July '42 - Dec '43) - Sobel was transferred to Regimental S-4 on 02DEC43.

2/LT-1/LT CLARENCE HESTER (ASN: O-425255) - Company HQ, XO (July '42 - Mar '43)

1/LT JAMES K. DAVIS (ASN: O-404800) - Company HQ, XO (Apr '43 - May '43) - Davis was transferred to Bn HQ in May 1943 to serve as HQ Company XO.

1/LT PATRICK SWEENEY (ASN: O-429890) - Company HQ, XO (Nov '43 - May '44) - Sweeney briefly served as the XO for E Co. after Sobel was transferred. He even had a short stint with E Co in March-May '43, including during part of the Tennessee Maneuvers.

SGT-1/SGT WILLIAM S. "STANTON" EVANS (ASN: 18117497) – Company 1st Sergeant

S/SGT MURRAY B. ROBERTS (ASN: 14083048) – Company HQ Supply Sgt - Roberts was briefly transferred to SVC/506 on 17AUG43. By October '43, he was back with E Co.

SGT ELMER L. "MOE" MURRAY JR (ASN: 19126297) – Company HQ Ops Sgt

PVT-SGT CARL N. RIGGS (ASN: 39103579) – Company HQ Comm Sgt - Riggs was promoted to the rank of T/5 on 20 Feb. 1943.

T/5 HERMAN F. "HERM" COLLINS (ASN: 11098666) – Company HQ - Collins was briefly transferred to SVC/506 on 17AUG43. He would return to E Co in time for D-Day.

T/5 JERRY A. WENTZEL (ASN: 19112534) – Company HQ Comm & 3rd Platoon - Wentzel was a member of 3rd Platoon before being transferred to Company HQ. Wentzel was promoted to PFC on 1 Feb 1943.

PVT-T/5 RALPH H. WIMER (ASN: 33446711) – Company HQ Medic

PFC GEORGE L. ELLIOT (ASN: 14139866) – Company HQ

PFC WILLIAM T. "BILL" MCGONIGAL (ASN: 11102116) - Company HQ Comm

PFC GERALD D. SNIDER (ASN: 15107810) - Company HQ Cook

PVT ERNEST L. OATS (ASN: 15331144) - Company HQ Medic

PVT JOSEPH "RED" HOGAN (ASN: 16035931) - Company HQ

PVT/PFC MAXWELL M. "MAX" CLARK (ASN: 11087989) - Company HQ Supply Clerk

Company Training Cadre

1/SGT HARVEY H. MOREHEAD (ASN: 6396532) - Company Cadre 1/SGT

S/SGT JOE E. STEDMAN (ASN: 6859578) - 1st Platoon Cadre SGT - Transferred to the 517th Prcht. Inf. On 15MAR43.

S/SGT NORMAN A. FORDE (ASN: 7032054) - 2nd Platoon Cadre SGT - Transferred to the 517th Prcht. Inf. On 15MAR43.

S/SGT STEVEN KUDLA (ASN: 13006511) - 3rd Platoon Cadre SGT - Transferred to the 517th Prcht. Inf. On 15MAR43.

1st Platoon

1st Platoon HQ

2/LT SALVE H. "MAT" MATHESON (ASN: O-444758) - 1st Platoon, Platoon Leader (July '42 - Nov '42)

2/LT-1/LT RICHARD D. "DICK" WINTERS (ASN: O-1286582) - Company HQ, XO, 1st Platoon & 2nd Platoon, Platoon Leader (2nd Plt: July '42 - May '43, XO: May '43 - Nov '43, 1st Plt: Nov '43 - June '44)

2/LT ARCHIBALD S. BARNWELL (ASN: O-447152) - 1st Platoon, Assistant Platoon Leader (Dec '42 - Apr '43) - Barnwell transferred out of E Co in April '43.

2/LT HARRY F. "LIMEY" WELSH (ASN: O-1318498) - 1st Platoon & 2nd Platoon, Assistant Platoon Leader (1st Plt: Apr '43 - June '44) - Welsh joined E Co at Ft. Bragg in April 1943.

PVT-S/SGT-PVT MYRON "MIKE" RANNEY (ASN: 17107816) - 1st Platoon, HQ & 2nd Squad

SGT LEO D. "FEARLESS PHOSGENE" BOYLE (ASN: 39315440) – 1st Platoon, Platoon Sergeant

PFC FRANK J. "PERCO" PERCONTE (ASN: 16100572) – 1st Platoon Comm - When Frank was asked if he was scared when he made his first jump, he said, "(Sure as) hell I was!" (Source: Muccia interview with Frank Perconte, undated)

T/5 GORDON "GORDY" CARSON (ASN: 12130792) – 1st Platoon Comm

T/4 GEORGE "PORTAGEE" LUZ (ASN: 11096443) – 1st Platoon Comm

PVT JOHN "JACK" MCGRATH (ASN: 32187910) – 1st Platoon Rockets

PVT ARTHUR J. "ART" MAUZERALL (ASN: 31326779) – 1st Platoon HQ

PFC JOSEPH "JOE" LIEBGOTT (ASN: 39103809) – 1st Platoon Comm

PVT EUGENE "GENE" ROE (ASN: 38265653) – 1st Platoon Medic

1st Squad, 1st Platoon

SGT ROBERT J. "ROOK" RADER (ASN: 20508327) – 1st Platoon, 1st Squad, Squad Leader

T/4 BURTON P. "PAT" CHRISTENSON (ASN: 19087456) – 1st Platoon, 1st Squad MG

PFC WOODROW W. "CURLY" ROBBINS (ASN: 39103557) – 1st Platoon, 1st Squad MG - Robbins was promoted to the rank of PFC on 28 Feb. 1943.

PFC WILLIAM "SHEP" HOWELL (ASN: 20508273) – 1st Platoon, 1st Squad

PVT RICHARD T. BRAY (ASN: 13106268) – 1st Platoon, 1st Squad

PVT ROBERT "RIP" VAN KLINKEN (ASN: 39186454) – 1st Platoon, 1st Squad - Van Klinken was one of the men who took demolition training. In addition to Wingett and Garrod, Rip used to pal around with "Skinny" Sisk of 3rd Platoon. (Source: Muccia interview with Bill Wingett, undated)

T/4 ROBERT B. "BURR" SMITH (ASN: 12131573) – 1st Platoon, 1st Squad/Company HQ Ops

PVT ROBERT J. "BOB" BLOSER (ASN: 13099636) – 1st Platoon, 1st Squad - Bloser was injured on his first jump.

PFC COBURN "JOHNNY" JOHNSON (ASN: 19138010) - 1st Platoon, 1st Squad

PFC EDWARD DONOHUE (ASN: 19087441) - 1st Platoon, 1st Squad

PFC JOHN P. SHEELEY (ASN: 39314111) - 1st Platoon, 1st Squad

CPL DENVER "BULL" RANDLEMAN (ASN: 16085544) - 1st Platoon, 1st Squad, Assistant Squad Leader

2nd Squad, 1st Platoon

SGT JOHN W. "JOHNNY" MARTIN (ASN: 15107902) - 1st Platoon, 2nd Squad, Squad Leader

PVT WILLIAM T. "BILL" WINGETT (ASN: 19087381) - 1st Platoon, 2nd Squad MG

PFC EVERETT J. GRAY (ASN: 19087506) - 1st Platoon, 2nd Squad MG - Gray had been teamed with Bill Wingett on a .30cal MG before Wingett was transferred to 2nd Bn HQ.

PVT ALBERT "ALBIE" BLITHE (ASN: 13111948) - 1st Platoon, 2nd Squad

PVT-CPL WILLIAM H. "DUKE" DUKEMAN (ASN: 19087501) - 1st Platoon, 2nd Squad, Assistant Squad Leader

PVT JACK F. MATTHEWS (ASN: 33051724) - 1st Platoon, 2nd Squad

PFC THOMAS A. "MACK" MCCREARY (ASN: 33150036) - 1st Platoon, 2nd Squad

PFC WILLIAM S. "CHICO' METZLER (ASN: 19126206) - 1st Platoon, 2nd Squad

PVT JAMES L. "JIM" SOWELL (ASN: 34802850) - 1st Platoon, 2nd Squad

PFC ROBERT K. MARSH (ASN: 35623704) - 1st Platoon, 2nd Squad - Marsh joined Easy Company at Fort Bragg before the unit shipped out for England.

CPL JAMES V. "RED" BENTON (ASN: 15107638) - 1st Platoon, 2nd Squad - Benton transferred to the 517th Airborne (D Company) on 15MAR43. Photographic evidence from the march from Toccoa to Atlanta suggest that Benton was likely in 2nd Squad as he is pictured near members of that squad which included Bill Wingett.

PFC DONALD B. "HOOBS" HOOBLER (ASN: 20508303) - 1st Platoon, 2nd Squad - Hoobler was promoted to PFC on 1 Feb 1943. He transferred to First Squad at some point.

PVT JACK O. GINN (ASN: 39099876) - 1st Platoon, 2nd Squad - Ginn was promoted to the rank of PFC on 28 Feb. 1943. Ginn went AWOL on 19 June 1943 and was reassigned to Bn HQ. From Tipper's anecdote, it was originally thought that Ginn had been assigned to 3rd Platoon, but photographic evidence suggests he was part of 1st Platoon.

Mortar Squad, 1st Platoon

PVT-SGT ROBERT T. "RT" SMITH (ASN: 19087366) - 1st Platoon Mortars, Mortar Squad Leader - Smith is listed on the Parachute Class #49 roster.

PVT EARL V. "DOPEY" BRUCE (ASN: 18076813) - 1st Platoon Mortars

PFC JOSEPH M. JORDAN (ASN: 15107761) - 1st Platoon Mortars

PFC ALEX M. "PINKY" PENKALA JR (ASN: 35549002) – 1st Platoon Mortars - Penkala ended up in Cook School in the Army.A couple of his buddies talked him into trying out for this new elite group called the Paratroopers. The story is that his friends washed out, but (he) became a member of the 506 PIR.

PVT RUDOLPH R. "BIG BOY" DITTRICH (ASN: 38459124) – 1st Platoon Mortars

PFC SERGIO G. MOYA (ASN: 39124765) – 1st Platoon Mortars/Company HQ Co Clerk - Moya entered the service in 1942 from South San Francisco. Prior to D-Day, Moya had been previously assigned to 1st Platoon's Mortar Squad.SGT Robert T. Smith, who was the Mortar Squad Leader at the time, stated that Moya "was the best damn paratrooper in the Company."

1st Platoon Unknown Squad/Section

PVT ROBERT "JEETER" LEONARD (ASN: 39115006) – 1st Platoon

PFC CLARENCE S. "TED" HOWELL (ASN: 34107814) – 1st Platoon

PVT WALTER "WALLY" WENTZEL (ASN: 16109311) – 1st Platoon

PVT KENNETH J. WEBB (ASN: 32383307) – 1st Platoon

PFC GEORGE P. YOCHUM (ASN: 13115745) – 1st Platoon - Yochum joined E Co at Ft. Benning on 8 Feb 1943 but was transferred on 25AUG43 to F Co while at Ft. Bragg. He harbored quite a bit of hatred for Captain Sobel for transferring him.

PFC JOHN N. "SMILING JACK" MILLER (ASN: 33216794) – 1st Platoon/Company HQ - Photographic evidence suggests that Jack Miller started out In 1st Platoon before transferring to Co HQ. In fact, a 1st Platoon roster dated 1 May 1944, compiled by 1/LT Winters, shows a "PVT. Miller" assigned to the platoon's 4th Squad.

PVT GEORGE J. RAJNER (ASN: 15107784) – 1st Platoon - Rajner was promoted to PFC on 1 Feb 1943. A machine gunner, Rajner was transferred to TPS (Assignment "X") on 13APR43. He served with the 505th PIR, 82nd Airborne and was KIA, 3 July 1944.

PFC - T/5 BENJAMIN J. "BUD" STONEY (ASN: 39530033) – 1st Platoon - Stoney was promoted to the rank of T/5 on 20 Feb. 1943.

PVT ROY E. PICKEL (ASN: 13125162) – 1st Platoon - Pickel started his training at Camp Croft, South Carolina and Ft. Benning, (jump school), from there he went to Camp Mackall and became a member of Easy Company. After training in Tennessee and Ft. Bragg he ended up in Camp Shanks. Finally, together with the others, he was shipped from New York to Liverpool on the RMS Samaria.

PVT VERLIN V. "SADIE" HAWKINS (ASN: 39395642) – 1st Platoon - Hawkins appeared on the 1942 2nd Battalion Christmas Roster. He later washed out of jump school for refusing to make his 5th jump and was transferred on 2 February 1943.

SGT TASKEL ELLIS (ASN: 12082030) – 1st Platoon - Ellis appeared on the 1942 2nd Battalion Christmas Roster. He later transferred to the 517th PIR on 15MAR43.

PFC RICHARD R. "DICK" GARROD (ASN: 39097956) – 1st Platoon, Toccoa - Garrod appeared on the 1942 2nd Battalion Christmas Roster. Garrod was transferred to Camp Shanks on 02SEP43. Garrod used to pal around with Wingett and Van Klinken.

PFC GEORGE E. JONES (ASN: UNK) – 1st Platoon - Jones appeared on the 1942 2nd Battalion Christmas Roster. Jones appeared on the 1942 2nd Battalion Christmas Roster.

S/SGT JACK B. SMOTHERS (ASN: 19126155) – 1st Platoon

PVT RICHARD F. BERG (ASN: 12131622) – 1st Platoon, Benning - Berg appeared on the 1942 2nd Battalion Christmas Roster. He was transferred in Feb 1943 to Casual Company, Parachute School, Ft. Benning, GA.

PVT JAMES F. "STINKY" COLEMAN (ASN: 13066282) – 1st Platoon, Benning - Coleman appeared on the 1942 2nd Battalion Christmas Roster. Transferred to TPS (Assignment "X") on 13APR43.

2nd Platoon

2nd Platoon HQ

2/LT WARREN R. "FARTSACK" ROUSH (ASN: O-1290664) – 2nd Platoon & 1st Platoon, Platoon Leader (1st Plt: Dec '42 – Feb '43, 2nd Plt: May '43 – June '44)

1/LT LEWIS "BLACKBEARD" NIXON (ASN: O-1286540) – 2nd Platoon, Assistant Platoon Leader (July '42 – Feb '43) - Nixon was promoted to 1LT on 16 Feb 1943.

S/SGT JAMES L. "PUNCHY" DIEL (ASN: 16100615) – 2nd Platoon, Platoon Sergeant

PFC RODERICK G. "ROD" BAIN (ASN: 19114638) – 2nd Platoon Comm

1st Squad, 2nd Platoon

SGT BERNARD "BARNEY" CUNNINGHAM (ASN: 39185303) – 2nd Platoon, 1st Squad Leader

PVT GEORGE L. POTTER JR (ASN: 19135776) – 2nd Platoon, 1st Squad - Potter joined the Company on 22 February 1943.

PVT EUGENE E. "JACK" JACKSON (ASN: 13011296) – 2nd Platoon

PVT EDWARD J. "EDDIE" JOINT (ASN: 13135150) – 2nd Platoon, 1st Squad - Ed was only 17 years old when he enlisted as a private with the Airborne. Ed took his basic training in Texas and afterwards he participated jump school training at Fort Benning, Georgia. (Source: Muccia interview with Ed Joint, undated)

PVT JAMES A. "MIKE" MCMAHON (ASN: 32815699) – 2nd Platoon, 1st Squad

PFC DEWITT "ALABAMA" LOWREY (ASN: 14131458) - 2nd Platoon MG

T/5 HERMAN "HANK" HANSON (ASN: 16100573) - 2nd Platoon, 1st Squad

PFC LEWIS N. "BOB" LAMPOS (ASN: 34571522) - 2nd Platoon, 1st Squad - Lampos was transferred to 2HQ/506 briefly on 25AUG43. He would return to E Co in time for D-Day.

CPL JAMES "JIM" CAMPBELL (ASN: 15107892) - 2nd Platoon, 1st Squad, Assistant Squad Leader - Campbell was promoted to the rank of PFC on 28 Feb. 1943.

2nd Squad, 2nd Platoon

SGT CHARLES "CHUCK" GRANT (ASN: 19126213) - 2nd Platoon, 2nd Squad Leader

PVT RICHARD D. "DICK" DAVENPORT (ASN: 32761716) - 2nd Platoon, 2nd Squad

PFC JOHN "JOHNNY" PLESHA (ASN: 39185305) - 2nd Platoon, 2nd Squad MG

PFC JOHN L. SHEEHY (ASN: 12131335) - 2nd Platoon, 2nd Squad

PVT-CPL-PVT CLEVELAND O. "O'NEILL" PETTY (ASN: 14078331) – 2nd Platoon, 2nd Squad MG - Petty was assigned to E Co on 8 Feb 1943 while at Ft. Benning. He had trained on machine guns prior to the transfer.

PVT EDWARD "EDDIE" MAUSER (ASN: 36312517) – 2nd Platoon, 2nd Squad MG (AG) - The U.S. Army drafted him on Jan. 17, 1942. At Fort Benning, Ga., he saw soldiers parachuting for the first time. "They called them paratroopers, and I thought, 'That looks exciting.'" Mauser's first Army outpost was with the horse cavalry at Fort Riley, Kan., which quickly transformed into a mechanized cavalry unit. Mauser was sent to Fort Benning, Ga., for reassignment when he saw paratroopers training. "That kind of fascinated me. So I volunteered," he said. Soldiers initially jumped while attached to harnesses and then progressed to jumping with parachutes from 30-foot, 45-foot and 250-foot towers. Each man was required to make five jumps from a C-47 during a week to earn his wings and the right to tuck his trousers into his boot tops for a bloused look. Mauser used rubber bands to achieve the paratrooper style. "When you're in the plane, you wonder what you're doing there," he said of training. "It was a little scary, but after it was over, it was fun." Mauser took his jump wings and moved on to Fort Bragg, Ga., where he joined the 101st Airborne and Easy Company. Mauser said his path to Easy Company bypassed the runs up Currahee Mountain and other excruciating physical training in Georgia at Camp Toccoa depicted in HBO's "Band of Brothers." But continued training at sites across the American southeast honed the elite fighting unit. Members jumped with full kits, including machine guns, mortars, explosives, grenades, maps, ammunition and rations. They were issued new weapons, clothing and jump gear. The men of Easy Company sailed out of New York Harbor and past the Statue of Liberty for Europe on Sept. 5, 1943. (Source: Hendee, David, Robb, DeDra, He Parachuted into History, World Herald, Omaha, NE, http://www.omaha.com/article/20090705/NEWS01/707059933)

PFC SALVATORE F. "SAL" BELLINO (ASN: 32681680) – 2nd Platoon, 2nd Squad

CPL JOSEPH D. "JOE" TOYE (ASN: 13026128) – 2nd Platoon, 2nd Squad, Assistant Squad Leader - Toye broke his leg and ankle on their 7th jump (30 Mar 1943).

PVT LEOPOLLO P. "ALEX" CARRILLO (ASN: 39857283) – 2nd Platoon, 2nd Squad - Carrillo appeared on the 1942 2nd Battalion Christmas Roster.

Mortar Squad, 2nd Platoon

SGT WILLIAM J. "WILD BILL" GUARNERE (ASN: 13113070) – 2nd Platoon Mortars, Mortar Squad Leader - "In mid-1942, July, I enlisted in the US Army paratroopers, and was shipped to Camp Toombs, GA for training. This there in GA, now called Toccoa, that I began what was to be the greatest adventure of my life. I was assigned to Co E 506 Parachute Regt, which was a part of the 101st Airborne Division. All the initial training called basic, was all physical conditioning. We trained, ran up Currahee Mountain in Toccoa, GA, for 6 months, then a forced march as made by 2nd BN 506, from Toccoa, GA, to Atlanta, GA, covering 125 miles. We made this march and set a record too. Still stands today. 3 days off to Ft. Benning, GA, for parachute training and packing chutes, mock up jumps and Dec 1942, and made 5 jumps and qualified for our coveted wings. They gave us a week furlough to go home, a proud moment in all of our young lives. We were paratroopers and could blouse our trousers in our jump boots; we thought we were the best. More training at Ft. Bragg, Mackall, and maneuvers in Tennessee 1943, then we shipped out in Spring 1943 on British ship Samaria, for Liverpool, England." (Source: Notes from William Guarnere, Band of Brothers Archival Material, Erik Jendresen Personal Collection, undated)

PFC WARREN H. "SKIP" MUCK (ASN: 12131169) – 1st Platoon/2nd Platoon Mortars

PFC DONALD G. "MALARK" MALARKEY (ASN: 39316320) – 2nd Platoon Mortars - Malarkey was promoted to the rank of PFC on 28 Feb. 1943.

PVT VERNON J. MENZE (ASN: 39040679) – 2nd Platoon Mortars

PVT ALTON "AL" MORE (ASN: 20949081) – 2nd Platoon Mortars - More joined the Company at Ft. Bragg.

PVT/CPL JAY S. DICKERSON (ASN: 14137803) – 2nd Platoon Mortars - Dickerson, who was friends with Don Malarkey and Skip Muck, was likely part of the Mortar Squad. He was removed from jump status and transferred on 18 Feb 1943 to Casual Company, Parachute School, Ft. Benning, GA.

2nd Platoon Unknown Squad/Section

PVT WALTER L. MCKAY (ASN: 36640669) – 2nd Platoon - Per the 13AUG43 morning report, McKay was reported as AWOL. The 14AUG43 morning reported that he was carried as such in error.

PVT J.B. STOKES (ASN: 38287058) – 2nd Platoon - Stokes joined E Co at Camp Mackall.

PFC THOMAS "TOMMY" BURGESS (ASN: 39185186) – 2nd Platoon

PVT WILLIAM D. "BILL" SERILLA (ASN: 16109180) – 2nd Platoon - Serilla is pictured in a pre-England group photo of mostly 2nd Platoon troopers, hence his placement in that platoon. He later transferred to the 504th PIR, 82nd ABN, sometime in 1943. Transferred to TPS (Assignment "X") on 13APR43. He had this to say in correspondence in the Sink Newsletter, "My airborne history began in Aug. 1942 after graduating from High School. I wanted to be a Marine Paratrooper but was informed they weren't training any, so I enlisted in the Army Paratroops. Went to Toccoa and assigned to E Co., 506. Took all the training including the 100-mile march from Toccoa to Atlanta in December (double time the last mile). Capt. Sobol (sic) was C.C. From Mackall, 100 men were selected (99 & me) and returned to the "Frying Pan Area", then to New York. Boarded the U.S. Washington with the 82nd A/B div. and went to Casablanca, North Africa. Jumped into Sicily, Italy and Holland. First night on Anzio Beach I captured a German Col."

PVT JOACHIM "JACKIE" MELO (ASN: 32820984) – 2nd Platoon

PVT GREGORY C. ROTELLA (ASN: 34036333) – 2nd Platoon

PVT FREDERICK G. BEALKE (ASN: 13102819) – 2nd Platoon - Bealke appeared on the 1942 2nd Battalion Christmas Roster.He was charged with being AWOL on 2 and 4 Feb 1943 and then transferred on 9 Feb 1943 to Casual Company, Parachute School, Ft. Benning, GA.

PVT EDWARD J. "RUSTY" BERNAT (ASN: 13089555) – 2nd Platoon - Ed Bernat and Eddie Joint met in jump school prior to being assigned to E/506.

PVT BRUCE A. HUDGENS (ASN: 16100406) – 2nd Platoon - Hudgens, originally from Chicago, IL, appeared on the 1942 2nd Battalion Christmas Roster. He also appeared in one of Don Malarkey's group photos. He joined E Co on 22AUG43 at Ft. Bragg after a stint at Lawson General Hospital in Atlanta.

3rd Platoon

3rd Platoon HQ

2/LT WALTER "WALLY" MOORE (ASN: O-454603) – 3rd Platoon, (July '42 – Feb '43)

2/LT GEORGE LAVENSON (ASN: O-1286510) – 3rd Platoon, Assistant Platoon Leader (Dec '42 – Mar '43) - Lavenson was one of the initial E company officers (with Winters, Nixon and Hester).

2/LT FREDERICK "MOOSE" HEYLIGER (ASN: O-1010235) - 3rd Platoon, Assistant Platoon Leader

PVT-S/SGT-PVT TERRENCE "SALTY" HARRIS (ASN: 19091437) - 3rd Platoon, Platoon Sergeant

PFC EDWIN E. "ED" PEPPING (ASN: 19127068) - 3rd Platoon Medic - After Pearl Harbor, Doc signed up for service in the Army in Los Angeles. He was processed into the service at Fort MacArthur. It was there that he spotted a recruiting table for the paratroops.There were several sharp dressed troopers with parachute insignia on their caps and on the desk they were sitting at was a sign that said, "Do you dare?" It was then that Doc decided he wanted to be a paratrooper. From Fort MacArthur, Doc took a train to Toccoa, Georgia for training. It was there he was introduced to Mount Currahee. It was a place he would come to know well, running the now famous mountain with the rest of the 506th Parachute Infantry Regiment.Once assigned to the 506th, he was detailed to join the Regiment's Medical Detachment. He would later be assigned to the 2nd Battalion's Company E, more commonly referred to as Easy Company. Once in Easy, Doc joined 3rd Platoon as the platoon medic. After parachute training at Fort Benning, Doc and the rest of the 506th took part in exercises in Tennessee, trained at bases such as Fort Bragg and Camp MacKall.They were finally moved up to Camp Shanks, New York in August of 1943. (Source: Ed Pepping, biographical notes, unpublished, via Matt Pepping, undated)

PVT - T/5 WILLIAM C. "BILL" MAYNARD (ASN: 39397626) – 3rd Platoon Comm - Maynard was promoted to the rank of PFC on 28 Feb. 1943.

1st Squad, 3rd Platoon

SGT CLIFFORD CARWOOD "LIP" LIPTON (ASN: 15116961) – 3rd Platoon, Platoon Sergeant & Squad Leader

PFC WALTER S. "SMOKEY" GORDON (ASN: 13099280) – 3rd Platoon, 1st Squad MG

PFC FRANCIS J. "FRANK" MELLETT (ASN: 20229437) – 3rd Platoon, 1st Squad MG (AG)

PVT - T/5 LEDLIE RIDLEY "PARK" PACE (ASN: 13099192) – 3rd Platoon, 1st Squad MG (AG) - Pace spent some time as Company Clerk and when he was transferred out of E Company Sobel considered making Tipper the clerk. Mellett took over as Gordon's AG when Pace was transferred. Pace was charged with being AWOL on 15 Feb 1943 and was taken off jump status on 19 Feb 1943. Transferred to TPS (Assignment "X") on 13APR43.

PVT - PFC EDWARD J. "ED" TIPPER (ASN: 16106128) – 3rd Platoon, 1st Squad Rockets - Tipper was promoted to the rank of PFC on 28 Feb. 1943.

PVT JOSEPH "JOE" RAMIREZ (ASN: 39103622) – 3rd Platoon, 1st Squad Rockets

PVT - SGT FLOYD "TAB" TALBERT (ASN: 15107832) – 3rd Platoon, Squad Leader - Talbert was promoted to the rank of T/5 on 20 Feb. 1943.

PFC FORREST L. "GOODIE" GUTH (ASN: 13099211) – 3rd Platoon, 1st Squad Grenadier, Company Armorer/Artificer - "We (Goodie, Strohl, and Fenstermaker) were in Third Platoon, and would all make it through the war! I earned the coveted jump wings on December 26, 1942. At the time we were still GHR (General Headquarters Reserve) troops and our only shoulder sleeve insignia was the corresponding patch (the 506th PIR was only attached to the 101st Airborne Division in June 1943). We finally prepared for movement overseas and embarked from NY port of embarkation with destination England; this was September 5, 1943. (Source: Muccia interview with Forrest Guth, undated)

PVT PHILIP "PHIL" PERUGINI (ASN: 12127453) – 3rd Platoon, 1st Squad

PVT ROBERT E. "POPEYE" WYNN (ASN: 13066265) – 3rd Platoon, 1st Squad

PVT WAYNE A. "SKINNY" SISK (ASN: 15116959) – 3rd Platoon, 1st Squad

PFC DANIEL B. "DB" WEST (ASN: 15116953) – 3rd Platoon, 1st Squad MG

PVT/PFC CARL L. "DUTCH" FENSTERMAKER (ASN: 13099212) – 3rd Platoon, 1st Squad

PVT RICHARD M. "RED" WRIGHT (ASN: UNK) – 3rd Platoon, 1st Squad

PVT WILLIAM "POP" WOODCOCK – 3rd Platoon, 1st Squad - Woodcock joined E Co at the Frying Pan at Ft. Benning in February 1943. Woodcock also served briefly as the Company Clerk at Camp Mackall.

PFC KENNETH T. BALDWIN (ASN: 13099345) – 3rd Platoon, 1st Squad - Baldwin was transferred to the 697th Field Artillery on 20AUG43. He was ultimately transferred to HQ, 2nd Bn, 501st PIR.

PVT URBAN M. SHIRLEY (ASN: 18109450) – 3rd Platoon, 1st Squad - Shirley appeared on the 1942 2nd Battalion Christmas Roster. Transferred to TPS (Assignment "X") on 13APR43.

PVT JOHN G. CONNELL (ASN: 13111956) – 3rd Platoon, 1st Squad - Connell appeared on the 1942 2nd Battalion Christmas Roster. Transferred to TPS (Assignment "X") on 13APR43. He would later be awarded a Silver Star while serving with the 82nd Airborne during the Battle of the Bulge. "The President of the United States of America, authorized by Act of Congress July 9, 1918, takes pleasure in presenting the Silver Star to Private First Class John G. Connell (ASN: 13111956), United States Army, for gallantry in action while serving with Company H, 3d Battalion, 505th Parachute Infantry Regiment, 82d Airborne Division, in action on 3 January 1945, in Belgium. Private First Class Connell, rifleman, was a member of the assault platoon. After seeing two of his comrades fall wounded from intense enemy machine gun fire, Private First Class Connell, with utter disregard for his own safety, crawled forward to a firing position and engaged six Germans armed with MP-43 weapons. He killed one, wounded two and caused the remaining three to surrender their positions. This allowed his squad to penetrate the enemy's main line of resistance and sever their communications. This soldier reflected great credit upon himself and the Airborne Forces." Clearly Connell was an excellent trooper and it was definitely Sobel's mistake in transferring him out of the company.

2nd Squad, 3rd Platoon

CPL-SGT AMOS J. "BUCK" TAYLOR (ASN: 13099363) – 3rd Platoon, Squad Leader - "I was a squad leader; I had 2nd Squad of the 3rd Platoon. Carwood Lipton was leader of the 1st Squad and Clarence Tridle was the mortar squad leader."

CPL ARTHUR "ART" YOUMAN (ASN: 39251325) – 3rd Platoon, Assistant Squad Leader

PVT-PFC WILLIAM F. "BILL" KIEHN (ASN: 19114027) – 3rd Platoon, 2nd Squad - Kiehn joined Easy while they were at Ft. Benning for parachute training. He is listed on the Parachute Class #49 roster.

PFC DARRELL C. "SHIFTY" POWERS (ASN: 13066266) – 3rd Platoon, 2nd Squad

PVT DONALD "DON" MOONE (ASN: 32772534) – 3rd Platoon, 2nd Squad - Don Moone joined E Co at Ft. Benning in June 1943.

PVT EARL J. "ONE LUNG" MCCLUNG (ASN: 39462287) – 3rd Platoon, 2nd Squad – Shifty had this to say about McClung: "Earl was one of the best. He was an excellent shot, and he could practically see in the dark, and weren't scared of anything." But when you got off the line, "Earl was a regular sad sack. I would press his clothes for him, get him out and lined up and make sure he passed inspection. Because if he didn't pass inspection, they would restrict us to the base for the weekend. So I took care of Earl cause I knew it would pay off when we got into combat and it did, it did. Just a top notch man, and good friend." Earl joined Easy at Ft. Bragg just before they left for England." (Source: Muccia interview with Darrell Powers, undated.)

Mortar Squad, 3rd Platoon

SGT CLARENCE M. "CRASH" TRIDLE (ASN: ASN: 17058836) – 3rd Platoon Mortars, Mortar Squad Leader

PFC RODERICK G. "ROD" STROHL (ASN: 13099213) – 3rd Platoon Mortars

PFC PAUL C. "HAYSEED" ROGERS (ASN: 17058810) – 3rd Platoon Mortars

PFC CAMPBELL T. "CT" SMITH (ASN: 39531483) – 3rd Platoon Mortars - Smith was promoted to the rank of PFC on 28 Feb. 1943.

PVT JAMES H. "MOE" ALLEY (ASN: 18166060) - 3rd Platoon Mortars

3rd Platoon Unknown Squad/Section

PFC ROBERT F. "BOB" MANN (ASN: 39185214) - 3rd Platoon

PVT WILLIAM "WILLIE" WAGNER (ASN: 13099382) - 3rd Platoon - He is listed on the Parachute Class #49 roster. He was reduced on 3 Feb 1943 from PFC to PVT for being AWOL.

PVT LAVON P. REESE (ASN: 15107830) - 3rd Platoon

PVT DAVID E. MORRIS (ASN: 34688410) - 3rd Platoon - Morris joined E Co on 29JUL43 at Ft. Bragg.

PVT JOHN L. "GEORGIA JAP" EUBANKS (ASN: 34252419) - 3rd Platoon

PFC ELMER N. SCHUYLER (ASN: 13023556) - 3rd Platoon

PFC CONRAD M. BAAY (ASN: 39186530) - 3rd Platoon - Baay was reported as AWOL on the 16AUG43 morning report. He was eventually transferred to the Regimental Guard House on 22AUG43.

CPL JOHN F. FIEGUTH (ASN: 39185220) – 3rd Platoon - Fieguth was transferred to Camp Shanks on 02SEP43.

Unknown Platoon

Many of the troopers listed here were either assigned to or reassigned from E Co while at Ft. Benning. They are identified by the word "Benning."

1/LT ALFRED J. "FROSTY" WINTER JR (ASN: O-308628) – Ft. Bragg - Winter was briefly carried on the E Co roster (including in May '43 during part of the Tennessee Maneuvers), was transferred to D Co on 26AUG43, and eventually finished his time with 2HQ/506.

2/LT GERALD M. SELLINGER (ASN: O-424540) – Ft. Bragg - Sellinger was transferred on 17AUG43 to 1st Special Troops, 2nd Army while at Ft. Bragg.

1/LT SAMUEL C. DEAN (ASN: O-347680) - Appeared on the Officer's MPR for May '43 as being dropped from the Regiment.

1/LT WILFRED B. JEFFERY (ASN: O-1284339) - Appeared on the Officer's MPR for May '43 as being dropped from the Regiment.

PVT BRUCE E. ALEXANDER (ASN: 32666850) – Camp Shanks - It's unclear when Alexander, a native of Albany, NY, joined E Co, but he was transferred out of the Company on 03SEP43 while at Camp Shanks.

SGT PAUL L. BECKER (ASN: 15107710) – Benning - Becker appeared on the 1942 2nd Battalion Christmas Roster. Transferred to TPS (Assignment "X") on 13APR43.

SGT-PVT SHERMAN M. IRISH (ASN: 39099887) – Benning - Irish was transferred to Camp Shanks on 02SEP43.

PVT-CPL RICHARD C. NORRIS (ASN: 35389735) – Camp Shanks

PFC JOHNNIE L. GATHINGS (ASN: 34307303)

PFC WALTER G. HOWARD (ASN: 34315821) – Benning - Howard appeared on the 1942 2nd Battalion Christmas Roster. He was assigned to Parachute Class #49 (December 14, 1942, to January 2, 1943).

PFC WARREN C. HUNTLEY (ASN: 34036001) – Benning - Huntley appeared on the 1942 2nd Battalion Christmas Roster. He was charged with being AWOL on 2 Feb 1943 (reduced from PFC to PVT). Transferred to the 517th Prcht. Inf. On 15MAR43.

PFC HENRY E. NELSON (ASN: 39393230) – Benning - Nelson appeared on the 1942 2nd Battalion Christmas Roster.

PFC JOHN W. ROSSMAN (ASN: 13089375) – Benning - Rossman appeared on the 1942 2nd Battalion Christmas Roster. He was charged with being AWOL on 2 Feb 1943 (reduced from PFC to PVT) and removed from jump status on 10 Feb.

PFC PAUL WAGNER (ASN: 39393314) – Benning - Wagner appeared on the 1942 2nd Battalion Christmas Roster. He was charged with being AWOL on 2 Feb 1943. Transferred to the 517th Prcht. Inf. On 15MAR43.

PFC PAUL J. SULLIVAN (ASN: 39314228) – Benning - Sullivan appeared on the 1942 2nd Battalion Christmas Roster.

PVT OWEN M. ANDREWS (ASN: 15107705) – Benning - Andrews appeared on the 1942 2nd Battalion Christmas Roster. He was assigned to Parachute Class #49 (December 14, 1942, to January 2, 1943).

PVT ORA M. CHILDERS (ASN: 18137423) – Benning - Childers appeared on the 1942 2nd Battalion Christmas Roster. He was assigned to Parachute Class #49 (December 14, 1942, to January 2, 1943).

PVT SAMUEL M. CAWTHON (ASN: 18109447) – Benning - Cawthon appeared on the 1942 2nd Battalion Christmas Roster. He was transferred on 9 Feb 1943 to the Parachute School, Ft. Benning, GA.

PVT CHESTER R. ESCHENBACH (ASN: 13091573) – Benning - Eschenbach appeared on the 1942 2nd Battalion Christmas Roster. He was transferred on 9 Feb 1943 to Casual Company, Parachute School, Ft. Benning, GA.

PVT LARRY G. GILES (ASN: 19138009) – Benning - Giles appeared on the 1942 2nd Battalion Christmas Roster. He was transferred on 20AUG43 to 1st Special Troops, 2nd Army while at Ft. Bragg.

PVT MILTON B. GLASS (ASN: UNK) – Benning - Glass appeared on the 1942 2nd Battalion Christmas Roster. He was removed from jump status and transferred On 18 Feb 1943 to Casual Company, Parachute School, Ft. Benning, GA.

PVT FRANK B. "COWBOY" GRANT (ASN: 39184975) – Benning - Grant appeared on the 1942 2nd Battalion Christmas Roster. Transferred to TPS (Assignment "X") on 13APR43.

PVT GENOA H. GRIFFITH (ASN: 7102065) – Benning

PVT ROBERT C. HENSLEY (ASN: 33726733) – Benning

PVT ORAN V. HOLBROOK (ASN: 39316313) – Benning - Holbrook appeared on the 1942 2nd Battalion Christmas Roster. He was transferred on 9 Feb 1943 to Casual Company, Parachute School, Ft. Benning, GA.

PVT DAVID L. HALTOM (ASN: 39396433) – Benning - Haltom appeared on the 1942 2nd Battalion Christmas Roster. He was transferred on 9 Feb 1943 to Casual Company, Parachute School, Ft. Benning, GA.

PVT ROBERT E. JARRETT (ASN: 14137845) – Benning - Jarrett appeared on the 1942 2nd Battalion Christmas Roster. He was transferred on 9 Feb 1943 to Casual Company, Parachute School, Ft. Benning, GA.

PVT ALBERT MAHMOOD (ASN: 17107651) – Benning - Mahmood appeared on the 1942 2nd Battalion Christmas Roster. He was transferred to D Company on 2 Feb. 1943. After reviewing the D/506 jump rosters, no "Mahmood" made that jump indicating that he was likely transferred again, possibly out of the paratroops.

PVT WALTER E. MARTIN (ASN: 16109322) – Benning - Martin appeared on the 1942 2nd Battalion Christmas Roster. Transferred to TPS (Assignment "X") on 13APR43.

PVT ROBERT MAXWELL (ASN: 15107683) – Benning - Maxwell appeared on the 1942 2nd Battalion Christmas Roster.

PVT GORDON L. NEUENFELDT (ASN: 17107823) – Benning - Neuefeldt appeared on the 1942 2nd Battalion Christmas Roster. He would be transferred out of E Co at Camp Shanks on 03SEP43.

PVT GORDON H. OIEN (ASN: 37550390) – Benning

PVT ALEX R. RACZKOWSKI (ASN: 36535694) – Benning - Raczkowski spoke fluent Polish.

PVT GARLAND R. SMITH (ASN: 35698915) – Benning - Smith appeared on the 1942 2nd Battalion Christmas Roster.

CPL-PVT JOSEPH STICKLEY (ASN: 6944370) – Benning - Stickley appeared on the 1942 2nd Battalion Christmas Roster. He was transferred on 9 Feb 1943 to Casual Company, Parachute School, Ft. Benning, GA.

PVT GEORGE W. THOMASON (ASN: 39530136) – Benning - Thomason appeared on the 1942 2nd Battalion Christmas Roster. He was transferred on 9 Feb 1943 to Casual Company, Parachute School, Ft. Benning, GA.

PVT JOSEPH P. WHITECAVAGE (ASN: 13100781) – Benning, Battalion Mess

PVT JERRY G. YOUNG (ASN: 39396382) - Benning - Young was assigned to Parachute Class #49 (December 14, 1942, to January 2, 1943).

PVT LAWRENCE HILLYER (ASN: 35442518) - Benning - Appeared on the Feb. 1943 morning report dated 3 Feb as being SIQ. He was transferred on 9 Feb 1943 to Casual Company, Parachute School, Ft. Benning, GA.

PVT HAROLD K. CRAMER (ASN: 35316504) - Benning - Cramer appears on the February 1943 morning report (15 Feb) as being assigned to the base hospital. He was transferred on 9 Feb 1943 to Casual Company, Parachute School, Ft. Benning, GA.

PVT FRANK J. KELEMAN (ASN: 33240053) - Benning - Was assigned to the company on 23 Feb. 1943 as seen in the Feb 1943 morning report.

PVT WILLIAM D. GIER (ASN: 39106833) - Benning - Gier was assigned to the company on 23 Feb. 1943 as seen in the Feb 1943 morning report. Transferred to TPS (Assignment "X") on 13APR43.

PVT LEON F. JACKSON (ASN: 18023046) - Camp Shanks - Jackson was transferred out of E Co on 03SEP43, while at Camp Shanks. Jackson was transferred to Camp Shanks on 02SEP43.

PVT RENE MARROQUIN (ASN: 39116528) – Camp Shanks - Marroquin was transferred out of E Co on 03SEP43, while at Camp Shanks. According to the 13AUG43 morning report, he had previously gone AWOL. He was eventually transferred to the stockade at the Presidio, CA on 25AUG43.

PVT RUDY I. MILTON (ASN: 6934978) – Camp Shanks - Milton was transferred out of E Co on 03SEP43, while at Camp Shanks.

PVT GUASIO "GUSSIE" MONROE (ASN: 32769306) – Camp Shanks - Monroe was transferred out of E Co on 03SEP43, while at Camp Shanks.

PVT CURTIS W. MOORE (ASN: 37007896) – Ft. Bragg - Moore appears on the 01AUG43 morning report while E Co was still at Ft. Bragg. He was later transferred to the 502nd Prcht Inf on 04AUG43.

PVT ROBERT E. BAILEY (ASN: 31175525) – Ft. Bragg - Bailey appears on the 01AUG43 morning report while E Co was still at Ft. Bragg. He was later transferred to the 697th Field Artillery on 20AUG43.

PVT THOMAS M. VAN THULL (ASN: 32607119) – Ft. Bragg - A native of Bergen County, New Jersey, Van Thull appeared on the 11AUG43 company morning report, stating that he was placed on detached service at Ft. Custer for an Airborne 'press conference'. On the September 1943 payroll roster there was a notation next to Van Thull's name that stated, "Confinement Awaiting Trial. Not to be paid."

PVT GEORGE H. ROBINSON (ASN: 6938237) – Ft. Bragg

PVT JOHN J. MCKINSTRY (ASN: 32821137) – Ft. Bragg - McKinstry was charged with being AWOL on 17AUG43. He was later transferred to 3HQ/506 on 25AUG43.

PVT OTTO A. PORATH (ASN: 36280005) – Ft. Bragg - Porath was transferred on 20AUG43 to 1st Special Troops, 2nd Army while at Ft. Bragg.

PVT WALTER MALATICH (ASN: 17077517) – Benning - Malatich appears on the 20AUG43 morning report. He later served in F Company. Malatich was transferred to Camp Shanks on 02SEP43.

PVT HARRY D. TICE JR. (ASN: 33321254) – Ft. Bragg - Tice was transferred on 20AUG43 to 1st Special Troops, 2nd Army while at Ft. Bragg.

PFC ANTHONY N. GUARNIERI (ASN: 32680795) – Ft. Bragg

PVT LUTHER J. GUILLORY (ASN: 18170804) – Ft. Bragg - Guillory was transferred to the 697th Field Artillery on 20AUG43.

PVT ALFRED E. FOGLE (ASN: 325966) – Ft. Bragg - Fogle was transferred to Camp Shanks on 02SEP43.

PVT EUGENE R. ELLIOT (ASN: 16131713) – Benning - Transferred to TPS (Assignment "X") on 13APR43.

PVT VERNON F. KNIGHT (ASN: 39317596) – Benning – Knight was assigned to the company on 23 Feb. 1943 as seen in the Feb 1943 morning report. Transferred to TPS (Assignment "X") on 13APR43.

Appendix III: GI Slang

AWOL - Absent Without Leave

Baby carriage — A cart used to carry heavy machine gun. These were extensively used by paratroopers, aka the M3A4 Utility Hand Cart.

Battery acid — Coffee, also known as ink, blackout, and paint remover.

Bazooka - M1 Rocket Launcher

Carbine - M1 Carbine or the paratrooper model, M1A1 Carbine

Chickenshit—Pointless, unnecessary attention to military detail.

Chow hound — A soldier that is a heavy eater, or always looking to eat.

C Ration – Canned food that included meat hash, meat and vegetables, pork and beans, and similar stable food.

D Ration – Army issued chocolate 'survival' bar manufactured by Hershey.

Dog biscuits – Large, thick, shelf-stable crackers that were issued to the men as part of their field rations. These were much like the hard tack issued to Civil War troops in terms of consistency.

Dog and Pony Show – A display or demonstration put on for senior officers.

DZ – A drop zone

Fart bag (or sack) – The issue sleeping bag

Feather merchant — A civilian working for the military.

Field, in the — A field problem (aka exercise) or campaigning against an enemy under actual or assumed conditions.

Furlough – Extended leave (see below), equivalent to a civilian vacation.

Garand – M1 Garand service rifle

G.I. — Government issue; while applied broadly to any piece of issued equipment, it became universally applied to the soldiers themselves. It was applied to any soldier who strictly followed the rules, i.e. "He is very G.I."

Goldbrick — Soldier with a talent for avoiding disagreeable work.

Grease Gun - M3 Submachine Gun

Guard house lawyer — A person who knows little but talks much about regulations, military law, and a soldier's rights.

Hard Boiled - Someone who is tough and set in their ways.

Hash mark — A service stripe on the sleeve of the Class A uniform which indicated years of service.

Hitting the silk — Conducting a parachute jump.

Incoming Mail - Enemy artillery

K.P. - Kitchen police aka kitchen duty, oftentimes used as a punishment.

Leave - To be granted excused absence from duty for a short period of time, aka 'being on pass.'

Maggie's drawers — Red flag used on rifle range to indicate a miss.

Meat Can - A mess tin, made of metal, and shaped like a oval bowl (with a levered latch), it had a top and came with a set of utensils. This is what soldiers typically eat their rations in, especially in the field.

Meat Wagon — An army ambulance

Mess Gear — A soldier's individual mess kit, knife, fork, spoon, and cup. See "Meat Can." For eating in the chow hall, soldiers would use a mess tray vice the issued meat can.'

Molar mechanic - A dentist

Noncom — A non-commissioned officer, aka corporals, sergeants, staff sergeants, technical sergeants and first sergeants. Also known as an NCO.

O.D. — Olive drab, or officer of the day.

On the carpet — Called before the commanding officer for disciplinary reasons.

Operator - Someone skilled in betting, taking bets, or shamming/goldbricking.

Over the hill — To desert, aka to go AWOL.

Pecker checker — A medic

Pill roller - A medic

Repple depple — A replacement depot

Shave tail — A second lieutenant

Short arm exam—Checking for VD (usually in a surprise, pre-revile "raid" by medics).

Skivvies - underwear

*SNAFU/snafu (pronounced snaffoo) — A completely confused situation, short for "situation normal, all f**ked up."*

Shit-On-a-Single or S.O.S. — Chipped or creamed beef on toast.

Sub-gun - Typically used to describe the M1 Thompson Submachine Gun, i.e. Thompson Sub-gun.

Top sergeant or top kick — The first sergeant

Walkie-Talkie — A small, portable two-way field radio, aka a 'handy talkie.'

XO - Executive Officer, or the officer that was second-in-command of a unit.

Bibliography

Interviews

The following Easy Company veterans were interviewed by the author for this volume.

James Alley, Roderick Bain, William Guarnere, Forrest Guth, Ed Joint, Don Malarkey, Al Mampre, Ed Mauser, Bill Maynard, Earl McClung, Ed Pepping, Frank Perconte, Phil Perugini, Roy Pickel, Darrell Powers, Paul Rogers, Rod Strohl, Amos Taylor, Ed Tipper, Bill Wingett

The following 506th Parachute Infantry Regiment veterans were interviewed by the author for this volume.

Charles Craton Jr., Richard Falvey, Ken Johnson, Paul Miller, Bill True

The author would like to thank the families of the following Easy Company veterans for their assistance while researching and writing this volume.

Albert Blithe, Gordon Carson, Burton Christenson, Norman Dittrich, William Dukeman, William Evans, Carl Fenstermaker, Walter Gordon, Herman Hanson, Frederick Heyliger, Joseph Jordan, George Lavenson, Joseph Liebgott, Carwood Lipton, George Luz, Warren Muck, Alex Penkala, George Potter, Myron Ranney, Lavon Reese, Eugene Roe, Robert Burr Smith, Herbert Sobel, Floyd Talbert, Joseph Toye, Robert Van Klinken, David Webster, Daniel West, Richard Wright, Robert Wynn

The author would like to thank the families of the following 506th Parachute Infantry Regiment veterans for their assistance while researching and writing this volume.

James "Pee Wee" Martin, David Moellenkamp, Robert Sink, Robert Strayer

Unpublished Sources

Alley, James, Interview conducted by Major M. Nichols, USA (Ret), undated

Ambrose, Stephen E., Easy Company Interview Collection, undated, (supplied to the author by Martin Morgan): Clarence Hester, C. Carwood Lipton, Don Malarkey, John Martin, Salve Matheson, Earl McClung, Darrell Powers, Wayne Sisk, Amos Taylor, Joe Toye, Harry Welsh, Dick Winters

Garcia, Hector, Letter to Tom Potter concerning Sergio Moya, 15 March, 2003

Malarkey, Don, Collection of Correspondence with Muccia, undated

Malarkey, Don, Collection of Papers, provided to the author by Chris Langlois via the Malarkey Family

McGrath, John, Interview conducted by Phillip G. Russo, October 2011

Pepping, Ed, biographical notes, unpublished, via Matt Pepping, undated

Tipper, Ed, Letter to Cathy West Landis concerning her father, Dan West, 17 November 2009

Libraries and Archives

George Koskimaki Collection, U.S. Army Heritage and Education Center (USAHEC), 950 Soldiers Drive, Carlisle, PA 17013

Richard D. Winters Collection, 1930-2005, U.S. Army Heritage and Education Center (USAHEC), 950 Soldiers Drive, Carlisle, PA 17013

Other Unpublished Sources

American Veterans Center, Band of Brothers Symposium Notes, 2007

Band of Brothers Archival Material, Erik Jendresen Personal Collection, undated. This collection included information on the following veterans: James Alley, Roderick Bain, Leo Boyle, Thomas Burgess, Burton Christenson, Walter Gordon, William Guarnere, Herman Hanson, Clarence Hester, C. Carwood Lipton, Don Malarkey, Don Moone, Walter Moore, Lewis Nixon, Darrell Powers, Robert Rader, Denver Randleman, Myron Ranney, Robert Burr Smith, Robert Strayer, Roderick Strohl, Floyd Talbert, Ed Tipper, Joseph Toye, David Webster, Harry Welsh, Richard D. Winters, Richard Wright, Robert Wynn

Smith, Robert Burr, "One Last Look Back", 1992

Academic and Military Research Papers

506th Parachute Infantry Regiment, Monthly Personnel Report - Officers, dated August 1942 to September 1943

506th Parachute Infantry Regiment, Payroll Report, dated August 1942 to September 1943

Carwood Lipton, Army Service Experience Questionnaire, Department of the Army, U.S. Army Military History Institute, Carlisle Barracks, PA, undated.

E Company, 506th Parachute Infantry Regiment, Morning Reports, dated August 1942 to September 1943

Published Sources

Books

Ambrose, Stephen E., Band of Brothers: E Company, 506th Regiment, 101st Airborne from Normandy to Hitler's Eagle's Nest, Simon & Schuster, New York, NY, 1992

Brotherton, Marcus, A Company of Heroes: Personal Memories about the Real Band of Brothers and the Legacy They Left Us, Penguin Publishing Group, New York, NY, May 3, 2011

Brotherton, Marcus, The Men of Easy Company, We Who Remain and Are Alive: Untold Stories from the Band of Brothers, Penguin Publishing Group, New York, NY, May 4, 2010

Brotherton, Marcus, Shifty's War: The Authorized Biography of Sergeant Darrell "Shifty" Powers, the Legendary Sharpshooter from the Band of Brothers, Dutton Caliber, New York, NY, May 1, 2012

Currahee Scrapbook, 506th Parachute Infantry Regiment, 20 July 1942 - 4 July 1945, Published in 1945

deTrez, Michael, The Way We Were: Cpl. Forrest Guth: 'E' Company, 506th PIR, 101st Airborne Division (WWII American Paratroopers Portrait Series, No. 1), D-Day Publishing, Normandy, France, 2004

Guanere, Wild Bill, Heffron, Edward "Babe", Post, Robyn, Brothers in Battle, Best of Friends: Two WWII Paratroopers from the Original Band of Brothers Tell Their Story, Dutton Caliber, New York, NY, October 7, 2008

Malarkey, Don, Welch, Bob, Easy Company Soldier: The Legendary Battles of a Sergeant from World War II's "Band of Brothers", St. Martin's Griffin, New York, NY, May 12, 2009

Rapport Jr., Leonard, Norwood, Arthur, Rendezvous with Destiny, House of Print, Madelja, MN, 1948

Webster, David Kenyon, Parachute Infantry: An American Paratrooper's Memoir of D-Day and the Fall of the Third Reich, Dell Publishing, New York, NY, March 1, 2008

Winters, Major Dick, Kingseed, Cole Col. USA (ret.), Beyond Band of Brothers: The War Memoirs of Major Dick Winters, Dutton Caliber, New York, NY, May 6, 2008

Periodicals and Newspaper Articles

Alexander, Larry, 'Brothers' vet recalls WWII, Intelligencer Journal, Lancaster, PA, 15 November 2007

America in WWII Magazine, Band of Brothers Special Edition, Harrisburg, PA, Spring, 2015

Anderson, Christopher J., Dick Winters' Reflections on His Band of Brothers, D-Day and Leadership, American History Magazine, Beverly Hills, CA, Jun 12, 2006, https://www.historynet.com/dick-winters-reflections-on-the-band-of-brothers-d-day-and-leadership/

Baer, Reid, November interview with Darrell "Shifty" Powers, November 2007, http://www.menstuff.org/columns/overboard/powers.html (dead link)

Beck, Ken, Maneuvers help win war on Western Front, Murfreesboro Post, 20 February 2011, https://rutherfordtnhistory.org/maneuvers-help-win-war-on-western-front/

Borrelli, Christopher, A Soldier's Tale, The Toledo Blade, Toledo, OH, 14 October 2001

Boule, Margie, Story of a Brother Killed in WWII Isn't Just Hollywood's, The Oregonian, Portland, OR, April 11, 2002

Cahill, Robert, Powers remembers 'Band of Brothers', Smyth County News, Wytheville, VA, 15 November 2006

"Close Up of the Jap Fighting Man", Infantry Journal, Reader's Digest: Volume 41, No.247, November 1942

Edwards, Jean, Vet's story fit for miniseries, Chicago Tribune, Chicago, IL, 8 November 2007

Feehan, Jim, They were the men of Easy Company, Skagit Valley Herald, Skagit County, WA, 5 September 2001

Fooks, John, Local Man Relied on Training and Sharp Wits to Survive Battles of World War II, Texarkana Gazette, Texarkana, TX, 19 December 1992

Gerjevic, Sandi, Forever Brothers, Anchorage Daily News, Anchorage, AK, 5 June 2001

Hendee, David, Robb, DeDra, He Parachuted into History, World Herald, Omaha, NE, http://www.omaha.com/article/20090705/NEWS01/707059933

Jones, Josh, Hometown Hero: Marshall Alumnus Clifford Carwood Lipton inspires D-Day mini-series "Band of Brothers", The Parthenon, Huntington, WV, 2 November 2007

Kashatus, William C., Toughest of the tough, The Citizen's Voice, Wilkes-Barre, PA, 29 May 2011

Lopez, Adam, Daughter of WWII veteran recalls father's influence, The Northern Star, Dekalb, IL, 30 November 2009

ParaDice Magazine, 506th Parachute Infantry Regiment, Toccoa, GA, March 1943

Purinton, Cait, HBO miniseries premiere spurs D-Day memories, The Kansas City Star, Kansas City, MO, 10 July 2001

Reiter, Jon, A WWII 'brotherhood' of Men, publisher and date unknown

Salvatori, Nico, Last 'Band of Brothers' WWII veteran from Erie dies, Erie Times News, Erie, PA, 5 June 2016

Torpy, Bill, Veteran avoids his Band of Brothers, The Atlanta Constitution Journal, Atlanta, GA, 9 November 2003

Uncredited, Training in Parachute Infantry, Geneva Daily Times, Geneva, NY, 9 September 1942

Walker, John, Army & Navy - Envelopment from the Sky, Time Magazine, 28 June 1943, http://content.time.com/-time/subscriber/article/0,33009,790990,00.html

Weyrich, Carleta, Hoobler fights in WWII elite 'Easy Company', The People's Defender, West Union, OH, April 2007

World War II Magazine, Band of Brothers Special Edition

Other Media

Dewitt Lowrey Interview, YouTube, https://www.youtube.com/watch?time_continue=3&v=PP90DzxyWSU

Washington, DC Band of Brothers Event Panel, WWII Foundation, 20 July 2011, https://www.youtube.com/watch?v=mjK8H-2-Vfc

Websites

McMahan, Andrew, Tennessee Maneuvers, 23 September 2019, The Tennessee State Library and Archives Blog, https://tslablog.blogspot.com/2019/09/tennessee-maneuvers.html

S.S./RMS Samaria information obtained at the following website:
https://www.norwayheritage.com/p_ship.asp?sh=samas

The Minnesota Historical Society,
https://www.mnhs.org/webarchive

https://www.timesunion.com/history/article/camp-shanks-rockland-county-world-war-ii-18382543.php

https://historynet.com/camp-shanks-last-stop-u-s-a/

Notes

FROM EVERY CORNER OF THE NATION

1 History, A Legacy of Leadership, Camp Toccoa at Currahee, Inc., undated, accessed online at http://camptoccoaatcurrahee.org/history.html

2 Ibid.

3 Notes from Robert Strayer, Band of Brothers Archival Material, Erik Jendresen Personal Collection, undated.

4 Anderson, Christopher J., Dick Winters' Reflections on His Band of Brothers, D-Day and Leadership, American History Magazine, Beverly Hills, CA, Jun 12, 2006, https://www.historynet.com/dick-winters-reflections-on-the-band-of-brothers-d-day-and-leadership/

5 Edwards, Jean, Vet's story fit for miniseries, *Chicago Tribune*, Chicago, IL, 8 November 2007

6 Added pay for enlisted paratroopers was $50.00. Officer paratroopers received $100.00. Perconte was enlisted so he actually received $50.00, not $100.00 each month.

7 Dewitt Lowrey Interview, YouTube, https://www.youtube.com/watch?time_continue=3&v=PP90DzxyWSU

8 Interview with Amos "Buck" Taylor from WWII Magazine, accessed online at http://www.menofeasycompany.com/home/index.php?page_id=33&link=20 (dead link), author copy.

9 Christopher Borrelli, Former Toledoan helped inspire HBO's 'Band of Brothers', *The Toledo Blade*, October 14, 2001, accessed online at http://www.toledoblade.com/Movies/2001/10/14/Former-Toledoan-helped-inspire-HBO-s-Band-of-Brothers.html#VXd3b6AASPGDmqmg.99

10 "It was the challenge. It was the difficulty that they described of getting into the paratroopers. Their physical condition had to be excellent; they had to be able to pass all these endurance tests. Then there was the equipment and the training. I remember they said the equipment to put a paratrooper in combat cost close to $10,000 which I think was

an exaggeration but that's what it said. They just built it up as an elite fighting unit, and it fired my imagination. I thought if it was tough to get into, I wanted to see what it would be like. That's the reason I went for the paratroops." Notes from C. Carwood Lipton, Band of Brothers Archival Material, Erik Jendresen Personal Collection, undated.

11 Carwood Lipton, Army Service Experience Questionnaire, Department of the Army, U.S. Army Military History Institute, Carlisle Barracks, PA, undated.

12 West's middle name was Berdette. According to Ed Tipper, the men would frequently call West, "Daniel Boone" due to his first and middle initials being "D.B."

13 Jim Feehan, They were the men of Easy Company, *Skagit Valley Herald*, September 5, 2001.

14 Muccia interview with James Alley, undated

15 Those thirteen troopers were: Winters, Heyliger, Shifty Powers, Popeye Wynn, Clarence Tridle, Forrest Guth, Buck Taylor, Carl Fenstermaker, Rod Strohl, Joseph Whitecavage, Walter Gordon, Paul Rogers, and Myron Ranney.

16 Memories of C. Carwood Lipton, Stephen Ambrose interview. Courtesy of the Richard D. Winters Collection, U.S. Army Heritage and Education Center, Carlisle, PA.

17 Boule, Margie, Story of a Brother Killed in WWII Isn't Just Hollywood's, *The Oregonian*, Portland, OR, April 11, 2002

18 Muccia interview with Ed Tipper, undated

19 Memories of John W. Martin, Stephen Ambrose interview. Courtesy of the Richard D. Winters Collection, U.S. Army Heritage and Education Center, Carlisle, PA.

20 William C. Kashatus, Toughest of the Tough, *The Citizens Voice*, May 29, 2011, accessed online at http://citizensvoice.com/arts-living/toughest-of-the-tough-1.1153337#ixzz1NpOlHMs6

21 Memories of Joe Toye, Stephen Ambrose interview. Courtesy of the Richard D. Winters Collection, U.S. Army Heritage and Education Center, Carlisle, PA.

22 Memories of Leo Boyle, Stephen Ambrose interview. Courtesy of the Richard D. Winters Collection, U.S. Army Heritage and Education Center, Carlisle, PA.

23 Robert Cahill, Powers remembers 'Band of Brothers', *Smyth County News*, November 15, 2006

24 Muccia interview with Darrell Powers, undated

25 Memories of Walter Gordon, Stephen Ambrose interview. Courtesy of the Richard D. Winters Collection, U.S. Army Heritage and Education Center, Carlisle, PA.

26 Marcus Brotherton, *We Who Are Alive and Remain: Untold Stories from the Band of Brothers*, Berkley Trade, May 4, 2010

27 Memories of Robert Rader, Stephen Ambrose interview. Courtesy of the Richard D. Winters Collection, U.S. Army Heritage and Education Center, Carlisle, PA.

28 Burgess was born in Decatur and grew up in Humboldt, Illinois. After graduating from high school, he went to Centralia, Washington to visit his brother who was working for the Long Bell Lumber Company. Burgess was still living in Washington state when he enlisted in the Army. Originally, he wanted to join the Air Corps because he wanted to learn to fly planes, but eventually he chose the paratroops.

29 Sandi Gerjevic, Forever Brothers, *Anchorage Daily News*, Anchorage, AK, June 5, 2001

30 American Veterans Center, Band of Brothers Symposium Notes, 2007

31 Malarkey, Don, Collection of Papers, provided to the author by Chris Langlois via the Malarkey Family

32 Memories of Walter Gordon, Stephen Ambrose interview. Courtesy of the Richard D. Winters Collection, U.S. Army Heritage and Education Center, Carlisle, PA.

33 Larry Alexander, "Brothers" Vet Recalls WWII, *Intelligencer Journal*, Lancaster, PA, 15 November 2007

34 Muccia interview with Forrest Guth, undated

35 Interview with Amos "Buck" Taylor from WWII Magazine, accessed online at http://www.menofeasycompany.com/home/index.php?page_id=33&link=20 (dead link), author copy.

36 Muccia interview with Bill Wingett, undated

37 Anderson, Christopher J., Dick Winters' Reflections on His Band of Brothers, D-Day and Leadership, American History Magazine, Beverly Hills, CA, Jun 12, 2006, https://www.historynet.com/dick-winters-reflections-on-the-band-of-brothers-d-day-and-leadership/

38 Interview with Amos "Buck" Taylor from WWII Magazine, accessed online at http://www.menofeasycompany.com/home/index.php?page_id=33&link=20 (dead link), author copy.

39 Memories of Robert T. Smith, Stephen Ambrose interview. Courtesy of the Richard D. Winters Collection, U.S. Army Heritage and Education Center, Carlisle, PA.

40 Memories of Robert Rader, Stephen Ambrose interview. Courtesy of the Richard D. Winters Collection, U.S. Army Heritage and Education Center, Carlisle, PA.

41 Robert Burr Smith, *"One Last Look Back"*, Excerpts from the unpublished writings of Robert Burr Smith regarding his World War II experiences, courtesy of Susan Smith Finn, originally written in 1982.

42 Robert Cahill, Powers remembers 'Band of Brothers', *Smyth County News*, November 15, 2006

43 Memories of John W. Martin, Stephen Ambrose interview. Courtesy of the Richard D. Winters Collection, U.S. Army Heritage and Education Center, Carlisle, PA.

44 According to Carwod, he overcame having flat feet due to all the marching and physical activity they were subjected to over time. It essentially forced him to develop arches in his feet. Notes from C. Carwood Lipton, Band of Brothers Archival Material, Erik Jendresen Personal Collection, undated.

45 Memories of Burton "Pat" Christenson, Stephen Ambrose interview. Courtesy of the Richard D. Winters Collection, U.S. Army Heritage and Education Center, Carlisle, PA.

46 Guarnere letter to Mike Ranney, 20 September 1983

47 Muccia interview with Guarnere, undated

48 I realize that Don's experience in W Company is slightly out of order due to his earlier mentioning in the paragraphs covering his arrival at E Company. Despite this small chronological faux pax, I felt that the narrative flowed smoothly in this manner vice shoehorning Don's E Co welcome later in the chapter. I hope the reader won't mind. Muccia interview with Don Malarkey, undated.

49 Malarkey, Don, Collection of Papers, provided to the author by Chris Langlois via the Malarkey Family

50 Don eventually ran into Gene again while on pass in Toccoa. Happy-go-lucky Don spotted Gene and called out "Hi, Gene!" Brown, a lieutenant

and therefore superior, pulled Don aside and called him out over his slip up. While he appreciated seeing a familiar face, Brown had career aspirations, and openly fraternizing with an enlisted man would be frowned upon by senior officers should they see it occur. For a detailed accounting of this incident, I recommend Don's own book, Easy Company Soldier, written with Bob Welch. Interestingly, this would not be the last time the pair would run into each other. The next time was a much less forgiving environment, the fighting for the town of Foy during the Battle of the Bulge.

"THAT DAMN MOUNTAIN"

51 Band of Brothers, American Veterans Center, 2012, accessed online at http://www.americanveteranscenter.org/wwiichronicles/wwii-chronicles-issue-xxxvii/the-band-of-brothers-e-company-506th-parachute-infantry-regiment-101st-airborne-division/

52 Muccia interview with Don Malarkey, undated

53 Muccia interview with Ed Tipper, undated

54 Memories of Richard D. Winters, Stephen Ambrose interview. Courtesy of the Richard D. Winters Collection, U.S. Army Heritage and Education Center, Carlisle, PA.

55 Memories of Burton "Pat" Christenson, Stephen Ambrose interview. Courtesy of the Richard D. Winters Collection, U.S. Army Heritage and Education Center, Carlisle, PA.

56 Muccia interview with Paul Rogers, undated

57 Purinton, Cait, HBO miniseries premiere spurs D-Day memories, *The Kansas City Star*, Kansas City, MO, 10 July 2001

58 Notes from Walter Gordon, Band of Brothers Archival Material, Erik Jendresen Personal Collection, undated.

59 Of the incident, Howell had this to say, "Hitting the bushes going up, but getting caught...had to make another trip up and down." Memories of William Howell, Stephen Ambrose interview. Courtesy of the Richard D. Winters Collection, U.S. Army Heritage and Education Center, Carlisle, PA.

Robert Rader also recalled this incident as well. Notes from Robert Rader, Band of Brothers Archival Material, Erik Jendresen Personal Collection, undated.

60 ParaDice Magazine, 506th Parachute Infantry Regiment, Toccoa, GA, March 1943

61 Robert Burr Smith, *"One Last Look Back"*, Excerpts from the unpublished writings of Robert Burr Smith regarding his World War II experiences, courtesy of Susan Smith Finn, originally written in 1982.

62 Interview with Frank Perconte from WWII Magazine, accessed online at http://www.menofeasycompany.com/home/index.php?page_id=27&link=20 (dead link), author copy.

63 Correspondence between David Campbell and Carwood Lipton, Research Note for Band of Brothers, dated 21 September 1999. Campbell furnished the author with a copy of the letter.

64 Memories of Clarence Hester, Stephen Ambrose interview. Courtesy of the Richard D. Winters Collection, U.S. Army Heritage and Education Center, Carlisle, PA.

65 Memories of Salve Matheson, Stephen Ambrose interview. Courtesy of the Richard D. Winters Collection, U.S. Army Heritage and Education Center, Carlisle, PA.

66 Marcus Brotherton, *We Who Are Alive and Remain: Untold Stories from the Band of Brothers*, Berkley Trade, May 4, 2010.

67 Memories of Forrest Guth and Rod Strohl, Stephen Ambrose interview. Courtesy of the Richard D. Winters Collection, U.S. Army Heritage and Education Center, Carlisle, PA.

SOBEL LEAVES HIS INDELIBLE MARK

68 Memories of Walter Gordon, Stephen Ambrose interview. Courtesy of the Richard D. Winters Collection, U.S. Army Heritage and Education Center, Carlisle, PA.

69 Memories of Robert Rader, Stephen Ambrose interview. Courtesy of the Richard D. Winters Collection, U.S. Army Heritage and Education Center, Carlisle, PA.

70 Eric Deggans, Joining the tough ranks of Easy Company, *St. Petersburg Times*, published September 7, 2001

71 Memories of Richard D. Winters, Stephen Ambrose interview. Courtesy of the Richard D. Winters Collection, U.S. Army Heritage and Education Center, Carlisle, PA.

72 Author's email exchange with Larry Alexander, undated

73 Memories of Robert Rader, Stephen Ambrose interview. Courtesy of the Richard D. Winters Collection, U.S. Army Heritage and Education Center, Carlisle, PA.

74 Memories of Walter Gordon, Stephen Ambrose interview. Courtesy of the Richard D. Winters Collection, U.S. Army Heritage and Education Center, Carlisle, PA.

75 Muccia interview with Don Malarkey, undated

76 The Sisk/Train story has been told and retold many times. Tony Coulter, a longtime friend of Bill Guarnere and Babe Heffron, was present for a luncheon with Major Winters, Babe, Jack Agnew, Forrest and Harriett Guth, and several other veterans and friends of the veterans, when he heard the story told. There was a lot of laughter and a lot of faces that were red with embarrassment. It was Moe Alley that told me the story during a phone interview. In a letter to Major Winters, Sisk obliquely referred to the incident. Floyd Talbert did so as well. It had obviously made its rounds with the veterans. Muccia interview with James Alley, undated.

77 Stephen Ambrose, Interview with Wayne Sisk, undated. Interview transcript courtesy of Martin Morgan.

78 Stephen Ambrose, Interview with Joe Toye, undated. Interview transcript courtesy of Martin Morgan.

79 Stephen Ambrose, Interview with John W. Martin, undated. Interview transcript courtesy of Martin Morgan.

80 Memories of Robert Rader, Stephen Ambrose interview. Courtesy of the Richard D. Winters Collection, U.S. Army Heritage and Education Center, Carlisle, PA.

81 Rader's insights into Sobel are particularly telling, especially how he described Sobel as taking the "responsibility on his shoulders." For Sobel, command of Easy Company was an all-consuming endeavor, one that he devoted himself to. Winters observed that Sobel didn't associate with other officers and had no real friends. That couldn't have been easy for the man to deal with. He simply lacked the camaraderie that helped the others get through the difficult times. In short, all the pressure was on him and he had no "release valve" as it were.

82 Interview with Amos "Buck" Taylor from WWII Magazine, undated, accessed online at http://www.menofeasycompany.com/home/index.php?page_id=33&link=20 (dead link), author copy.

83 Stephen Ambrose, Interview with Amos Taylor, undated. Interview transcript courtesy of Martin Morgan.

84 Chris Anderson, *"A Pillar of Easy Company: George Luz"*, Stephen Ambrose Tours, 29 Decemebr 2017, accessed online at: https://stephenambrosetours.com/pillar-easy-company-george-luz/

85 Reid Baer, *"A Real Man from the Band of Brothers"* - Darrell Powers' – Remembrances, 2007

86 Angela Broyles, Dewitt Lowrey, World War II Veteran; Original Member of Band of Brothers; Historical Truth 101; 8 February 2010; accessed online at: http://historicaltruth101.com/2013/10/12/dewitt-lowery/; this article misspells Lowrey's last name and has several mistakes in it.

87 Memories of Robert Wynn, Stephen Ambrose interview. Courtesy of the Richard D. Winters Collection, U.S. Army Heritage and Education Center, Carlisle, PA.

88 Memories of Walter Gordon, Stephen Ambrose interview. Courtesy of the Richard D. Winters Collection, U.S. Army Heritage and Education Center, Carlisle, PA.

89 Band of Brothers, American Veterans Center, 2012, accessed online at http://www.americanveteranscenter.org/wwiichronicles/wwii-chronicles-issue-xxxvii/the-band-of-brothers-e-company-506th-parachute-infantry-regiment-101st-airborne-division/

90 Ambrose, Stephen E., Band of Brothers: E Company, 506th Regiment, 101st Airborne from Normandy to Hitler's Eagle's Nest, Simon & Schuster, New York, NY, 1992

91 Stephen Ambrose, Interview with Joe Toye, undated. Interview transcript courtesy of Martin Morgan.

SINK'S MUSCLE COLLEGE

92 Stephen Ambrose, Interview with John W. Martin, undated. Interview transcript courtesy of Martin Morgan.

93 Muccia interview with Ed Tipper, undated

94 Don was slightly off in this quote. They did quite a bit of military training as this was their "basic training" and as such, they were instructed in the rudiments of soldiering. Many of the other Easy men recalled weapons classes, drill and ceremony (ie, rifle manual of arms and marching in formation), map and compass work, bayonet training, as detailed below.

Stephen Ambrose, Interview with Don Malarkey, undated. Interview transcript courtesy of Martin Morgan.

95 Memories of C. Carwood Lipton, Stephen Ambrose interview. Courtesy of the Richard D. Winters Collection, U.S. Army Heritage and Education Center, Carlisle, PA.

96 Memories of Walter Gordon, Stephen Ambrose interview. Courtesy of the Richard D. Winters Collection, U.S. Army Heritage and Education Center, Carlisle, PA.

97 Ibid.

98 Gordon's comment of "...without reservation or pity," is interesting as it was a common theme when talking to the veterans about going up against each other in training. They couldn't be seen by Sobel as slacking off and giving their opposite number a break or an easier go. If they did, punishment would surely follow. So, each went "full tilt" when training against other squads/platoons/companies. Another instance where this is reflected in this book was the portion detailing the Friday night boxing matches at Toccoa.

99 Muccia interview with Paul Rogers, undated

100 Notes from Burton "Pat" Christenson, Band of Brothers Archival Material, Erik Jendresen Personal Collection, undated.

101 Notes from Richard D. Winters, Band of Brothers Archival Material, Erik Jendresen Personal Collection, undated.

102 Muccia interview with Ed Tipper, undated.

103 It was Paul Rogers who told me about the barrel-swapping story and its result. I can still see Paul smiling broadly in my mind as he remembered his good friend's antics. I never did get to ask him where Gordon got the black paint to paint the barrel. Muccia interview with Paul Rogers, undated

104 Notes from Burton "Pat" Christenson, Band of Brothers Archival Material, Erik Jendresen Personal Collection, undated.

105 Memories of Robert T. Smith, Stephen Ambrose interview. Courtesy of the Richard D. Winters Collection, U.S. Army Heritage and Education Center, Carlisle, PA.

106 Larry Alexander, "Brothers" Vet Recalls WWII, *Intelligencer Journal*, Lancaster, PA, 15 November 2007

107 Stephen Ambrose, Interview with Darrell Powers, undated. Interview transcript courtesy of Martin Morgan.

108 Zeroing is typically done at 25 yards. You fire one round at a time and made sight adjustments to ensure accuracy. This process allowed the shooter to set his sights. Once that was completed, and the shooter knew what his setting was, he could manipulate the sight for different ranges to increase his probability of a hit.

109 Muccia interview with Al Mampre, undated

110 Muccia interview with Darrell Powers, undated

111 Muccia interview with Amos Taylor, undated

112 Notes from C. Carwood Lipton, Band of Brothers Archival Material, Erik Jendresen Personal Collection, undated.

113 Marcus Brotherton, *We Who Are Alive and Remain: Untold Stories from the Band of Brothers*, Berkley Trade, May 4, 2010

114 Being in a machine gun team together fostered a deep friendship between Wingett and Gray. The pair had to develop a synergy...an innate ability to work together without much being said...an intuitiveness as to what was needed to be done, whether it was swapping out the cloth ammunition belts or shifting fire to cover a different area. Wingett and Gray bonded and became close as brothers over time. Muccia interview with Bill Wingett, undated

115 Memories of Walter Gordon, Stephen Ambrose interview. Courtesy of the Richard D. Winters Collection, U.S. Army Heritage and Education Center, Carlisle, PA.

116 Notes from Burton "Pat" Christenson, Band of Brothers Archival Material, Erik Jendresen Personal Collection, undated.

117 Muccia interview with Don Malarkey, undated

118 Malarkey, Don, Collection of Papers, provided to the author by Chris Langlois via the Malarkey Family

119 Notes from Richard D. Winters, Band of Brothers Archival Material, Erik Jendresen Personal Collection, undated.

120 Brotherton, Marcus, Shifty's War: The Authorized Biography of Sergeant Darrell "Shifty" Powers, the Legendary Sharpshooter from the Band of Brothers, Dutton Caliber, New York, NY, May 1, 2012

121 When I spoke to Shifty about this incident, he told me that he swore that the five-dollar bill that Popeye threw into the helmet was the same five-

dollar bill Popeye had bet him about joining the paratroopers. He had no way of proving it, but he sure enjoyed believing that it was. Muccia interview with Darrell Powers, undated

122 Robert Cahill, Powers remembers 'Band of Brothers', *Smyth County News*, November 15, 2006

123 The "Band Aid Bandits" was a nickname that Mampre and Pepping bestowed upon themselves.

124 Muccia interview with Ed Pepping, undated

125 Muccia interview with Al Mampre, undated

126 Muccia interview with Don Malarkey, undated

127 Notes from Burton "Pat" Christenson, Band of Brothers Archival Material, Erik Jendresen Personal Collection, undated.

128 Muccia interview with Amos Taylor, undated

129 Muccia interview with Amos Taylor, undated

130 Notes from Walter Gordon, Band of Brothers Archival Material, Erik Jendresen Personal Collection, undated.

131 Notes from Richard D. Winters, Band of Brothers Archival Material, Erik Jendresen Personal Collection, undated.

132 Stephen Ambrose, Interview with Clarence Hester, undated. Interview transcript courtesy of Martin Morgan.

133 Notes from Richard D. Winters, Band of Brothers Archival Material, Erik Jendresen Personal Collection, undated.

134 Notes from Walter Gordon, Band of Brothers Archival Material, Erik Jendresen Personal Collection, undated.

135 Notes from Richard D. Winters, Band of Brothers Archival Material, Erik Jendresen Personal Collection, undated.

136 Memoirs of Burton "Pat" Christenson, The Richard D. Winters Collection, Army Heritage and Education Center, Carlisle Barracks, Carlisle, PA, undated.

137 Ibid.

138 Muccia interview with Ed Tipper, undated

139 Notes from Ed Tipper, Band of Brothers Archival Material, Erik Jendresen Personal Collection, undated.

140 Stephen Ambrose, Interview with Amos Taylor, undated. Interview transcript courtesy of Martin Morgan.

141 *Close Up of the Jap Fighting Man*, Infantry Journal, Reader's Digest: Volume 41, No.247, November 1942

142 Memories of Gordon Carson, Stephen Ambrose interview. Courtesy of the Richard D. Winters Collection, U.S. Army Heritage and Education Center, Carlisle, PA.

143 Stephen Ambrose, Interview with Carwood Lipton, undated. Interview transcript courtesy of Martin Morgan.

144 Notes from Ed Tipper, Band of Brothers Archival Material, Erik Jendresen Personal Collection, undated.

THE LAST STEP BEFORE JUMP SCHOOL

145 I understand that the ethnic pejorative in this anecdote is out of synch with current social norms, but because this is the way the men spoke during that period, I left the language intact from their interviews throughout the narrative. Notes from John W. Martin, Band of Brothers Archival Material, Erik Jendresen Personal Collection, undated.

146 Memories of Gordon Carson, The Richard D. Winters Collection, Army Heritage and Education Center, Carlisle Barracks, Carlisle, PA, undated.

147 According to Burr Smith, Stoney's other nickname was, "Bulldog."

148 The M1919 air-cooled .30 caliber light machine gun weighed 32 pounds, not 36 as stated in the quote from the article. Additionally, I quoted the article verbatim and as a result, left "machinegun" as a single word (vice the conventional usage which separates it into two words; machine and gun). *600 Singing Paratroopers Do 115-Mile March in Three Days*, Lamar Q. Ball, The Atlantic Journal-Constitution, 4 December 1942

149 William Torpy, Veteran Avoids his Band of Brothers, *The Atlanta Journal-Constitution*, 9 November 2003

150 Notes from Walter Gordon, Band of Brothers Archival Material, Erik Jendresen Personal Collection, undated.

151 Notes from Richard D. Winters, Band of Brothers Archival Material, Erik Jendresen Personal Collection, undated.

152 Memoirs of Roderick Bain, The Richard D. Winters Collection, Army Heritage and Education Center, Carlisle Barracks, Carlisle, PA, undated.

153 Notes from Forrest Guth, Band of Brothers Archival Material, Erik Jendresen Personal Collection, undated.

154 Memories of Gordon Carson, The Richard D. Winters Collection, Army Heritage and Education Center, Carlisle Barracks, Carlisle, PA, undated.

155 Notes from Joe Toye, Band of Brothers Archival Material, Erik Jendresen Personal Collection, undated.

156 Notes from Richard D. Winters, Band of Brothers Archival Material, Erik Jendresen Personal Collection, undated.

157 Malarkey, Don, Collection of Papers, provided to the author by Chris Langlois via the Malarkey Family

158 Muccia interview with Don Malarkey, undated

159 Interview with William Wingett from WWII Magazine, undated, accessed online at http://www.menofeasycompany.com/home/index.php?page_id=35&link=20 (dead link), author copy.

160 Purinton, Cait, HBO miniseries premiere spurs D-Day memories, *The Kansas City Star*, 10 July 2001

161 Rod told me that Winters didn't really want to punish him as he understood why Strohl had left formation, not for his own benefit, but for that of the men in his squad. But Sobel was always watching and if Winters didn't do it, he would, and then Winters would feel his wrath as well. Still, Rod was unfazed by the punishment. "I needed the rest anyway," he said of being restricted to camp while the others went into town. Muccia interview with Rod Strohl, undated

162 Notes from William Guarnere, Band of Brothers Archival Material, Erik Jendresen Personal Collection, undated.

163 600 Singing Paratroopers Do 115-Mile March in Three Days, Lamar Q. Ball, *The Atlantic Journal-Constitution*, 4 December 1942

164 Memoirs of Roderick Bain, The Richard D. Winters Collection, Army Heritage and Education Center, Carlisle Barracks, Carlisle, PA, undated.

PARACHUTE TRAINING – EARNING THOSE SILVER WINGS

165 Muccia interview with Don Malarkey, Tonawanda, NY, 2-4 July 2010.

166 Interview with William Wingett from WWII Magazine, undated, accessed online at http://www.menofeasycompany.com/home/index.php?page_id=35&link=20 (dead link), author copy.

167 "Hey kid. Ya wanna know what we used ta blouse our jump boots? Rubbahs!" This was how Wild Bill explained how they bloused their jump boots to me. We had been discussing the jump boots and when they were allowed to begin blousing them. I explained that we Marines, used elastic bands that you could buy at the post exchange. Bill, in his typical way, told

me that why would you waste the money on buying rubber bands when prophylactic kits were free at the Battalion Aid Station and generally kept in a glass bowl on the first sergeant's desk (where you would go to get your weekend pass...if Sobel didn't take it away for some infraction, real or perceived). Muccia interview with William Guarnere, undated

168 Notes from Robert Rader, Band of Brothers Archival Material, Erik Jendresen Personal Collection, undated.

169 Notes from Richard D. Winters, Band of Brothers Archival Material, Erik Jendresen Personal Collection, undated.

170 Memoirs of Burton Christenson, "Basic Training", The Richard D. Winters Collection, Army Heritage and Education Center, Carlisle Barracks, Carlisle, PA, undated

171 Muccia interview with Ed Pepping, undated

172 When I spoke with Phil about jump school, he basically said, "We had some fun with the instructors." I didn't know how much fun it was for Phil in particular until I spoke with Doc Pepping. Muccia interview with Phil Perugini, Hershey, PA, 21 March 2011.

173 Muccia interview with Bill Wingett, undated

174 Winters, Major Dick, Kingseed, Cole Col. USA (ret.), Beyond Band of Brothers: The War Memoirs of Major Dick Winters, Dutton Caliber, New York, NY, May 6, 2008

175 Notes from Joe Toye, Band of Brothers Archival Material, Erik Jendresen Personal Collection, undated.

176 Memoirs of Robert T. Smith, The Richard D. Winters Collection, Army Heritage and Education Center, Carlisle Barracks, Carlisle, PA, undated

177 Memoirs of Roderick Bain, The Richard D. Winters Collection, Army Heritage and Education Center, Carlisle Barracks, Carlisle, PA, undated.

178 Muccia interview with Amos Taylor, undated

179 Ibid

180 Memories of Gordon Carson, The Richard D. Winters Collection, Army Heritage and Education Center, Carlisle Barracks, Carlisle, PA, undated.

181 Robert Burr Smith, *"One Last Look Back"*, Excerpts from the unpublished writings of Robert Burr Smith regarding his World War II experiences, courtesy of Susan Smith Finn, originally written in 1982.

182 Muccia interview with Frank Perconte, undated

183 Stephen Ambrose, Interview with Darrell Powers, undated. Interview transcript courtesy of Martin Morgan.

184 Memoirs of Forrest Guth and Rod Strohl, The Richard D. Winters Collection, Army Heritage and Education Center, Carlisle Barracks, Carlisle, PA, undated

185 Robert Burr Smith, *"One Last Look Back"*, Excerpts from the unpublished writings of Robert Burr Smith regarding his World War II experiences, courtesy of Susan Smith Finn, originally written in 1982.

186 Memoirs of Forrest Guth and Rod Strohl, The Richard D. Winters Collection, Army Heritage and Education Center, Carlisle Barracks, Carlisle, PA, undated

187 Muccia interview with Darrell Powers, undated

188 Muccia interview with Forrest Guth, undated

189 Robert Burr Smith, *"One Last Look Back"*, Excerpts from the unpublished writings of Robert Burr Smith regarding his World War II experiences, courtesy of Susan Smith Finn, originally written in 1982.

EASY COMPANY IN TRANSITION

190 Notes from Robert Rader, Band of Brothers Archival Material, Erik Jendresen Personal Collection, undated.

191 Notes from C. Carwood Lipton, Band of Brothers Archival Material, Erik Jendresen Personal Collection, undated.

192 Memories of Gordon Carson, The Richard D. Winters Collection, Army Heritage and Education Center, Carlisle Barracks, Carlisle, PA, undated

193 Memoirs of Robert T. Smith, The Richard D. Winters Collection, Army Heritage and Education Center, Carlisle Barracks, Carlisle, PA, undated

194 Rotella, from Asheville, NC, was eventually assigned to 128th Infantry Regiment, 32nd Infantry Division. He was killed in action on February 9, 1944, near Saidor, New Guinea and is buried in the Manila American Cemetery.

195 Notes from C. Carwood Lipton, Band of Brothers Archival Material, Erik Jendresen Personal Collection, undated.

196 Muccia interview with Amos Taylor, undated

197 When I spoke with Bill about this incident, he struggled to remember Rotella's name. Exasperated, he looked over at Babe, who was also present, and said, "Babe, what was that kid's name? The one they kicked

out?" Babe, not missing a beat, piped back, "How the hell do I know? I wasn't there!" Everyone broke up laughing. Muccia interview with William Guarnere, undated.

198 Muccia interview with William Guarnere, undated

199 Although Freeman had not joined Easy by that time, he knew Randleman well despite the pair being in different platoons. Brad provided the insights about Bull in this paragraph. Muccia interview with Bradford Freeman, undated.

200 Memories of Gordon Carson, The Richard D. Winters Collection, Army Heritage and Education Center, Carlisle Barracks, Carlisle, PA, undated

BECOMING MASTERS OF THEIR CRAFT

201 Notes from C. Carwood Lipton, Band of Brothers Archival Material, Erik Jendresen Personal Collection, undated.

202 Memoirs of Burton Christenson, The Richard D. Winters Collection, Army Heritage and Education Center, Carlisle Barracks, Carlisle, PA, undated

203 Memoirs of Gordon Carson, The Richard D. Winters Collection, Army Heritage and Education Center, Carlisle Barracks, Carlisle, PA, undated

204 Fooks, John, "Local Man Relied on Training and Sharp Wits to Survive Battles of World War II", "Days Gone By", *Texarkana Gazette*, Texarkana, Texas, 19 December 1992.

205 Stephen Ambrose, Interview with Wayne Sisk, undated. Interview transcript courtesy of Martin Morgan.

206 Muccia interview with Ed Tipper, undated

207 Marcus Brotherton, *We Who Are Alive and Remain: Untold Stories from the Band of Brothers*, Berkley Trade, May 4, 2010.

208 Muccia interview with Ed Tipper, undated

209 Reiter, Jon, A WWII 'brotherhood' of Men, publisher and date unknown

210 Morehead survived World War II, and continued his service during the Korean War, where he was captured by the North Koreans and spent time as a POW before being repatriated.

211 Memoirs of Robert T. Smith, The Richard D. Winters Collection, Army Heritage and Education Center, Carlisle Barracks, Carlisle, PA, undated

212 Notes from Walter Gordon, Band of Brothers Archival Material, Erik Jendresen Personal Collection, undated.

213 Notes from Robert Rader, Band of Brothers Archival Material, Erik Jendresen Personal Collection, undated.

214 Memoirs of Gordon Carson, The Richard D. Winters Collection, Army Heritage and Education Center, Carlisle Barracks, Carlisle, PA, undated

215 Muccia interview with Al Mampre, undated

216 Notes from Robert Wynn, Band of Brothers Archival Material, Erik Jendresen Personal Collection, undated.

217 Ambrose, Stephen E., Band of Brothers: E Company, 506th Regiment, 101st Airborne from Normandy to Hitler's Eagle's Nest. New York: Simon & Schuster, 2004. Pages 35-36.

218 Muccia interview with James Alley, undated

219 Muccia interview with Paul Rogers, undated

220 Rogers and Alley were more direct about Tip's unwillingness to share his "ill-gotten gains." Ed said with a wry smile, "That's not true. I shared with them...just not all the time." Muccia interview with Ed Tipper, undated.

221 Ambrose, Stephen E., Band of Brothers: E Company, 506th Regiment, 101st Airborne from Normandy to Hitler's Eagle's Nest. New York: Simon & Schuster, 2004. Page 36.

222 Notes from Robert Rader, Band of Brothers Archival Material, Erik Jendresen Personal Collection, undated.

223 Muccia interview with Al Mampre, undated

224 Muccia interview with Ed Pepping, undated

225 While this story has appeared in multiple publications over the years, it was told to me first, by Al Mampre at an Easy Company reunion in Chicago. Al, who was about as good-natured human being as you will find, was particularly proud of his and Pepping's deviousness in this escapade. He was especially proud that Sobel was unable to prosecute them for what they did. A year or two later, I attended a Currahee Military Weekend in Toccoa, Georgia. Along with Al, Easy veterans Ed Pepping and Brad Freeman were in attendance. It was there that Al and Ed recounted to me the entire incident in full detail.

226 According to Carwood Lipton, Connell was shipped out to a regular infantry unit and eventually entered combat in North Africa.

227 Baer, Reid, November interview with Darrell "Shifty" Powers, November

2007, http://www.menstuff.org/columns/overboard/powers.html (dead link)

228 Colonel Robert Strayer "Poop Sheets", Letter from Strayer to Winters, dated 24 August 1990.

229 Muccia interview with Ed Joint, undated

230 Memoirs of Gordon Carson, The Richard D. Winters Collection, Army Heritage and Education Center, Carlisle Barracks, Carlisle, PA, undated

231 Memories of Wayne Sisk, The Richard D. Winters Collection, Army Heritage and Education Center, Carlisle Barracks, Carlisle, PA, undated

232 Memories of Gordon Carson, The Richard D. Winters Collection, Army Heritage and Education Center, Carlisle Barracks, Carlisle, PA, undated

233 Ibid.

234 Notes from Robert Rader, Band of Brothers Archival Material, Erik Jendresen Personal Collection, undated.

235 Ed once told me that he understood why some of his fellow veterans' opinions about Sobel had softened over the years, but for him it never did. He felt that Sobel was too personal in targeting him and that it crossed a line between doing his military duty and just being spiteful. That is why Ed said that his hatred for Sobel never lessened. Muccia interview with Ed Tipper, undated.

236 Notes from Robert Rader, Band of Brothers Archival Material, Erik Jendresen Personal Collection, undated.

237 Memories of Gordon Carson, The Richard D. Winters Collection, Army Heritage and Education Center, Carlisle Barracks, Carlisle, PA, undated

238 Malarkey, Donald, Welch, Bob, "Easy Company Soldier", St. Martin's Press, New York, 2008, pages 58-59.

239 Malarkey, Letter to Muccia, dated 20 June 2007. In the letter, Don explained that as the reserve chute billowed back into his face, he thought, "Well, I've done it now. I've had it." A shock wave of fear went through him as he realized that he might not survive this. So, landing in the tree was a pleasant surprise for him as one would imagine.

240 Muccia interview with Roy Pickel, undated

THE GRADUATION EXERCISE

241 History of the Maneuvers, The Tennessee Maneuvers: A Temporary Interruption, 2018-2021, https://tennesseemaneuvers.org/maneuvers/

242 Burns, Frank, Second Army (Tennessee) Maneuvers, Tennessee Historical Society, Tennessee Encyclopedia, October 8, 2017, https://tennesseeencyclopedia.net/entries/second-army-tennessee-maneuvers/

243 Caldwell, Durham, Tennessee Maneuvers of 1943, article undated, https://sites.rootsweb.com/~tnsumner/tid1943.htm

244 Wood, Sterling A., Colonel (USA), History of the 313th Infantry in World War II, Infantry Journal Press, Washington D.C., March, 1947.

245 Memoirs of Robert T. Smith, The Richard D. Winters Collection, Army Heritage and Education Center, Carlisle Barracks, Carlisle, PA, undated

246 Memoirs of Burton Christenson, The Richard D. Winters Collection, Army Heritage and Education Center, Carlisle Barracks, Carlisle, PA, undated

247 Muccia interview with Don Malarkey, undated

248 Horn, Stanley Jr., "Blue Paratroopers Capture Roads, Bridges Near Gallatin", The Tennessean, 09JUN43.

249 Memoirs of Joe Toye, The Richard D. Winters Collection, Army Heritage and Education Center, Carlisle Barracks, Carlisle, PA, undated

250 Horn, Stanley Jr., "Blue Paratroopers Capture Roads, Bridges Near Gallatin", The Tennessean, 09JUN43.

251 The colored chutes and their corresponding supplies were as follows: ammunition & explosives, red; rations, green; medical supplies, white; fuel, blue; communications equipment, yellow.

252 Horn, Stanley Jr., "Reds Rally After Surprise Attack, Capture Sky Raiders", The Tennessean, 10JUN43.

253 Memories of Gordon Carson, The Richard D. Winters Collection, Army Heritage and Education Center, Carlisle Barracks, Carlisle, PA, undated

254 Memoirs of Robert T. Smith, The Richard D. Winters Collection, Army Heritage and Education Center, Carlisle Barracks, Carlisle, PA, undated

255 Both Carson and Smith recalled the incident with Blithe vividly in their memoirs. Interestingly, Blithe faced no repercussions from failing to take part in the jump. This was likely because he wasn't the only one who got airsick.

256 Johnny's recollection was incorrect. The E Co morning report for that day indicated it was a daytime jump. This was corroborated by Gordy Carson's jump log.

257 Memoirs of John W. Martin, The Richard D. Winters Collection, Army Heritage and Education Center, Carlisle Barracks, Carlisle, PA, undated

258 Memoirs of Paul Rogers, The Richard D. Winters Collection, Army Heritage and Education Center, Carlisle Barracks, Carlisle, PA, undated

259 Muccia interview with Don Malarkey, undated

260 Walker, John, Army & Navy - Envelopment from the Sky, Time Magazine, 28 June 1943, http://content.time.com/time/subscriber/article/0,33009, 790990,00.html

261 Beck, Ken, Maneuvers help win war on Western Front, Murfreesboro Post, 20 February 2011, https://rutherfordtnhistory.org/maneuvers-help-win-war-on-western-front/

262 https://tslablog.blogspot.com/2019/09/tennessee-maneuvers.html

263 Ray Nagell's recollections appear courtesy of the Minnesota Historical Society, https://www.mnhs.org/mgg/artifact/nagelloh

264 Doyle Gaines' account appears courtesy of the following website, although it appears to be a dead link at the time of this book's publication. http://www.tn4me.org/article.cfm/a_id/220/minor_id/72/major_id/9/era_id/7

265 Muccia interview with Don Malarkey, undated

266 Memories of Gordon Carson, The Richard D. Winters Collection, Army Heritage and Education Center, Carlisle Barracks, Carlisle, PA, undated

267 Carwood Lipton, Army Service Experience Questionnaire, Department of the Army, U.S. Army Military History Institute, Carlisle Barracks, PA, undated.

FORT BRAGG TO CAMP SHANKS AND THEN...ENGLAND

268 Winters commented that the executive, "was surrounded by more beautiful, well-groomed secretaries than he'd ever seen in his life." A situation that made the shy young officer with a Mennonite background feel extremely awkward. Winters, Major Dick, Kingseed, Cole Col. USA (ret.), Beyond Band of Brothers: The War Memoirs of Major Dick Winters, Dutton Caliber, New York, NY, May 6, 2008

269 Winters, Major Dick, Kingseed, Cole Col. USA (ret.), Beyond Band of Brothers: The War Memoirs of Major Dick Winters, Dutton Caliber, New York, NY, May 6, 2008

270 Don Moone had this to say about joining the paratroops, "(I was) naïve. (I had a) good recruiter. (I saw the) wings, boots, extra pay...(I was) just plain stupid." Don Moone, The Richard D. Winters Collection, Army Heritage and Education Center, Carlisle Barracks, Carlisle, PA, undated

271 David Morris, The Richard D. Winters Collection, Army Heritage and Education Center, Carlisle Barracks, Carlisle, PA, undated

272 McGrath, John, Interview conducted by Phillip G. Russo, October 2011

273 Shifty was initially unsure about discussing Mellett as he felt it was akin to "speaking ill of the dead." But he eventually talked about his relationship with Frankie with me and later, Marcus Brotherton, who captured it in far better detail than I've outlined here. Muccia interview with Darrell Powers, undated

274 Memoirs of Robert T. Smith, The Richard D. Winters Collection, Army Heritage and Education Center, Carlisle Barracks, Carlisle, PA, undated

275 McGrath, John, Interview conducted by Phillip G. Russo, October 2011

276 It was Wild Bill who explained to me the reasoning behind sewing a silver dollar behind the parachute patch on their overseas cap. Muccia interview with William Guarnere, undated

277 David Morris, The Richard D. Winters Collection, Army Heritage and Education Center, Carlisle Barracks, Carlisle, PA, undated

278 Memoirs of Richard D. Winters, The Richard D. Winters Collection, Army Heritage and Education Center, Carlisle Barracks, Carlisle, PA, undated

279 Memoirs of Burton Christenson, The Richard D. Winters Collection, Army Heritage and Education Center, Carlisle Barracks, Carlisle, PA, undated

280 Ibid.

281 Ibid.

282 Memoirs of C. Carwood Lipton, The Richard D. Winters Collection, Army Heritage and Education Center, Carlisle Barracks, Carlisle, PA, undated

283 Memoirs of Burton Christenson, The Richard D. Winters Collection, Army Heritage and Education Center, Carlisle Barracks, Carlisle, PA, undated

284 Muccia interview with William Guarnere, undated

285 Muccia interview with Bill Wingett, undated

286 Memoirs of Burton Christenson, The Richard D. Winters Collection, Army Heritage and Education Center, Carlisle Barracks, Carlisle, PA, undated

287 Memoirs of Richard D. Winters, The Richard D. Winters Collection, Army Heritage and Education Center, Carlisle Barracks, Carlisle, PA, undated

288 Data regarding the S.S./RMS Samaria can be found at the following link: https://www.norwayheritage.com/p_ship.asp?sh=samas